Lecture Notes in Mathematics

Volume 2240

More information about this series at http://www.springer.com/series/304

Ameya Pitale

Siegel Modular Forms

A Classical and Representation-Theoretic
Approach

 Springer

Ameya Pitale
Mathematics
University of Oklahoma
Norman, OK, USA

ISSN 0075-8434   ISSN 1617-9692   (electronic)
Lecture Notes in Mathematics
ISBN 978-3-030-15674-9   ISBN 978-3-030-15675-6   (eBook)
https://doi.org/10.1007/978-3-030-15675-6

Library of Congress Control Number: 2019934367

Mathematics Subject Classification (2010): 11-XX, 11-02, 11Fxx, 20-XX, 20Cxx, 20Kxx

This Springer imprint is published by the registered company Springer Nature Switzerland AG
The registered company address is: Gewerbestrasse 11, 6330 Cham, Switzerland

Contents

Introduction

The theory of modular forms is an important topic of research in number theory. The modular forms are the centerpiece of several recent accomplishments such as the proof of the Shimura–Taniyama–Weil conjecture due to Andrew Wiles and others, which led to the resolution of Fermat's Last Theorem. Siegel modular forms are the higher dimensional analogues of modular forms and are the subject of these lecture notes. Siegel modular forms have seen tremendous success recently in both analytic and arithmetic aspects, with results toward Langlands functoriality and Deligne's conjectures.

These lecture notes are based on my workshop on Siegel modular forms at IISER, Pune, India from August 8 to 18, 2017. The target audience for the workshop, and for these notes, are graduate students and young researchers. These notes could also be used by senior researchers as reference material.

The main objective of the workshop was to introduce classical and representation theoretic techniques in modular forms and to explicate the interconnections via current research and open problems. I believe that the classical and representation theoretic methods are a two-way street and it is absolutely essential for researchers to gain expertise in going from one side to the other and back.

For this purpose, I decided to focus the workshop, and these lecture notes, on Siegel modular forms. These are higher dimensional analogues of elliptic modular forms and there is a long history of active research on this topic. The main advantage is that one can approach Siegel modular forms from a purely classical point of view as holomorphic functions on the Siegel upper half-space. In addition, these can be studied in the context of local and global representation theory of the symplectic group.

The subject of Siegel modular forms is vast and it is impossible to cover all of it in a 2-week workshop. This is my disclaimer that these notes are not comprehensive. I have tried to give a detailed description of the basics of the classical theory and representation theory of Siegel modular forms. But beyond the basics, my emphasis is more on how to use the concepts to solve several interesting problems and to give an indication about the current open questions in the subject. I have tried to provide references for anything that is not proved in the lecture notes so that the reader can access details. I hope that this will open certain locked doors for the readers, and get them excited enough to pursue learning or researching on these and related topics.

Some of the topics that have been omitted are the Galois representations associated to Siegel modular forms, the paramodular conjecture, and vector-valued Siegel modular forms. The first two omissions are by choice since I do not work on problems related to them, while the last one is just to simplify things by restricting to scalar-valued Siegel modular forms. Each of these aspects of Siegel modular forms could be one or more chapters by themselves. Instead, I am just going to provide some references to get the interested readers started on them. For Galois representations look at [57] and [105], for the paramodular conjecture see [16] and for vector-valued Siegel modular forms refer to [71].

There are several good texts which explain details of the classical theory of Siegel modular forms (see [4, 28, 46, 59]) and some research articles that provide details of the representation theory of the symplectic group ([6, 81]). There is a gap in the literature when it comes to a good text or reference article giving the necessary in-depth information on the two approaches to Siegel modular forms as well as the interconnections between the two. I hope that these notes will fill that gap.

An important feature of the notes is that I have tried to provide several exercises. These play a dual role. They allow me to introduce certain results without having to go through their proofs. But more importantly, anyone wanting to really understand the material can only do so if they get their hands dirty doing these exercises. For the convenience of the reader, I have provided solutions (or hints) for all the exercises in an appendix.

In an ideal situation, the reader will already be familiar with the $GL(2)$ theory. This will help in realizing how the theory and methods for Siegel modular forms are often a generalization of the elliptic modular forms theory. Nevertheless, I have included three appendices recalling the classical concepts and representation theory for $GL(2)$ and basics of p-adic numbers and the ring of adeles.

The notes are organized as follows. Chapters 1–3 introduce basic information on the classical theory of Siegel modular forms. Chapters 4 and 5 introduce advanced topics, open conjectures and recent results that use the techniques introduced in the preceding Chapters. Chapter 6 is the transition from the classical to the representation theoretic. Chapter 7 provides much of the basic local representation theory. Chapters 8–10 present current research and introduce the various techniques and concepts required to understand them.

Acknowledgements: I would like to thank everyone at IISER, Pune, and especially Baskar Balasubramanyam, for their hospitality and seamless organization. I am also grateful to the participants of the workshop who spent close to 30 hours with me over a period of 10 days, and countless many hours working on the exercises by themselves. Their feedback was crucial in correcting mistakes and typos in earlier versions of the notes. I would also like to mention that the workshop was part of the Global Initiative of Academic Networks (GIAN) proposal of the government of India.

I am thankful for all the help and guidance provided by Ralf Schmidt in making sure that the lecture notes are accurate. I am also grateful to the referees whose comments and suggestions were invaluable.

Finally, I want to thank my family—Swapna, Aayush, and Samay—for the support and encouragement that makes all of it possible. These lecture notes are dedicated to them.

Chapter 1
Introduction to Siegel Modular Forms

In this chapter, we will begin with a problem of representation of integers or matrices by quadratic forms. This acts as a starting point to the introduction of Siegel modular forms. We will then discuss the symplectic group and its action on the Siegel upper half-space. Finally, the definition and basic properties of Siegel modular forms will be introduced. Good references for this material are books by Andrianov [4], Freitag [28], Klingen [46] and Maass [59].

1.1 Motivation

Let m be a positive integer. Let $A \in M_m(\mathbb{Z})$ be such that

- A is even, i.e., all the diagonal entries of A are even,
- A is symmetric, i.e., $A = {}^t A$, the transpose of A, and,
- $A > 0$, i.e., positive definite.

Define,

$$Q(x) := \frac{1}{2} {}^t x A x = \sum_{1 \le i < j \le m} a_{ij} x_i x_j + \sum_{i=1}^{m} \frac{1}{2} a_{ii} x_i^2, \qquad x \in \mathbb{R}^m,$$

an integral, positive definite quadratic form in variables $x_1, x_2, \cdots, x_m$. For $n \in \mathbb{N}$, set

$$r_Q(n) := \#\{x \in \mathbb{Z}^m : Q(x) = n\}$$

the number of representations of n by Q.

Exercise 1.1 *Using that A is positive definite, show that $r_Q(n) < \infty$.*

Problem: *Find an exact formula for $r_Q(n)$, or at least asymptotics for $r_Q(n)$ as $n \to \infty$.*

© Springer Nature Switzerland AG 2019
A. Pitale, *Siegel Modular Forms*, Lecture Notes in Mathematics 2240,
https://doi.org/10.1007/978-3-030-15675-6_1

Jacobi's idea: Consider the theta series

$$\Theta_Q(z) := 1 + \sum_{n \geq 1} r_Q(n)e^{2\pi i n z} = \sum_{x \in \mathbb{Z}^m} e^{2\pi i Q(x)z}, \qquad z \in \mathbb{H}_1 = \{x + iy \in \mathbb{C} : y > 0\}.$$

Let N be a positive integer such that NA^{-1} is also an even, integral matrix. Suppose that m is even. By Corollary 4.9.5 (iii) of [62], it is known that $\Theta_Q \in M_{m/2}(N) =$ space of modular forms of weight $m/2$ and level N. Essentially,

$$\Theta_Q(\frac{az+b}{cz+d}) = \pm(cz+d)^{m/2}\Theta_Q(z), \qquad \begin{bmatrix} a & b \\ c & d \end{bmatrix} \in SL_2(\mathbb{Z}), \, N|c.$$

Now, use the fact that $M_{m/2}(N)$ is a finite-dimensional $\mathbb{C}$-vector space to get results for $r_Q(n)$. This is illustrated in the following exercise.

Exercise 1.2 *Let* $m = 4$, $Q(x) = x_1^2 + x_2^2 + x_3^2 + x_4^2$. *Then, we know that* $\Theta_Q \in M_2(4)$. *The dimension of* $M_2(4)$ *is 2, and has basis* $E_1(z) = P(z) - 4P(4z)$, *and* $E_2(z) = P(z) - 2P(2z)$, *where*

$$P(z) = 1 - 24\sum_{n \geq 1} \sigma_1(n)e^{2\pi i n z}, \text{ with } \sigma_1(n) = \sum_{d|n} d.$$

Use this information to find a formula for $r_Q(n)$ *and, as a corollary, obtain Lagrange's theorem that* $r_Q(n) \geq 1$, *whenever* $n \geq 1$.

More general problem: *Study the number of representations of a* $n \times n$ *matrix* T *by* Q, *i.e.*,

$$r_Q(T) := \#\{G \in M_{m,n}(\mathbb{Z}) : \frac{1}{2} {}^t G A G = T\}. \tag{1.1}$$

Observe that, for $r_Q(T) > 0$, we need $T = {}^t T$, $T \geq 0$ and T is half-integral, i.e., $2T$ is even. Put

$$\Theta_Q^{(n)}(Z) := \sum_{\substack{T = {}^t T \geq 0 \\ T \text{ half integral}}} r_Q(T)e^{2\pi i \operatorname{Tr}(TZ)}, \qquad Z \in \mathbb{H}_n.$$

$\mathbb{H}_n$ is the Siegel upper half-space defined in (1.4). For m even, Satz A2.8 of [28] says that $\Theta_Q^{(n)}$ is a *Siegel modular form* of weight $m/2$, level N and degree n. Here, N is again defined as in the previous modular forms case. We will next describe the symplectic group and define Siegel modular forms.

1.2 The Symplectic Group

Let $n \in \mathbb{N}$ and R be a commutative ring with 1. The *symplectic group of similitudes* is defined by

$$\mathrm{GSp}_{2n}(R) := \{g \in \mathrm{GL}_{2n}(R) : {}^t g J g = \mu(g) J, \mu(g) \in R^\times, J = \begin{bmatrix} 0_n & 1_n \\ -1_n & 0_n \end{bmatrix}\}.$$

The function $\mu : \mathrm{GSp}_{2n}(R) \to R^\times$ is called the multiplier (homomorphism). We have the subgroup

$$\mathrm{Sp}_{2n}(R) := \{g \in \mathrm{GSp}_{2n}(R) : \mu(g) = 1\}.$$

Exercise 1.3 *Let* $g = \begin{bmatrix} A & B \\ C & D \end{bmatrix} \in \mathrm{GL}_{2n}(R)$, *with* $A, B, C, D \in M_n(R)$. *Then the following are equivalent.*

(i) $g \in \mathrm{GSp}_{2n}(R)$ *with multiplier* $\mu(g) = \mu$.
(ii) ${}^t g \in \mathrm{GSp}_{2n}(R)$ *with multiplier* $\mu({}^t g) = \mu$.

(iii) $\mu g^{-1} = \begin{bmatrix} {}^t D & -{}^t B \\ -{}^t C & {}^t A \end{bmatrix}$.

(iv) *The blocks* A, B, C, D *satisfy the conditions*

$$ {}^t A C = {}^t C A, {}^t B D = {}^t D B \text{ and } {}^t A D - {}^t C B = \mu 1_n. \tag{1.2}$$

(v) *The blocks* A, B, C, D *satisfy the conditions*

$$ A\,{}^t B = B\,{}^t A, C\,{}^t D = D\,{}^t C \text{ and } A\,{}^t D - B\,{}^t C = \mu 1_n. \tag{1.3}$$

Note that this immediately implies that $\mathrm{GSp}_2(R) = \mathrm{GL}_2(R)$.
Examples of symplectic matrices: We have the following matrices in $\mathrm{Sp}_{2n}(R)$.

(i) $\begin{bmatrix} 1_n & X \\ & 1_n \end{bmatrix}$, where $X = {}^t X \in M_n(R)$.

(ii) $\begin{bmatrix} g & \\ & {}^t g^{-1} \end{bmatrix}$, where $g \in \mathrm{GL}_n(R)$.

(iii) $\begin{bmatrix} & 1_n \\ -1_n & \end{bmatrix}$.

(iv) $K_n := \{\begin{bmatrix} X & Y \\ -Y & X \end{bmatrix} : X, Y \in M_n(R), X\,{}^t Y = Y\,{}^t X, X\,{}^t X + Y\,{}^t Y = 1_n\}$.

Exercise 1.4 *The Iwasawa decomposition states that we can write any $g \in \mathrm{Sp}_{2n}(\mathbb{R})$ as*

$$g = \begin{bmatrix} 1_n & X \\ & 1_n \end{bmatrix} \begin{bmatrix} g & \\ & {}^t g^{-1} \end{bmatrix} k, \qquad X = {}^t X \in M_n(\mathbb{R}), g \in \mathrm{GL}_n(\mathbb{R}), k \in K_n.$$

Use this to show that $\mathrm{Sp}_{2n}(\mathbb{R}) \subset \mathrm{SL}_{2n}(\mathbb{R})$.

Over an arbitrary ring R, one can use induction to show that $\mathrm{Sp}_{2n}(R)$ is generated by matrices of the form (i), (ii), and (iii) above. This immediately gives $\mathrm{Sp}_{2n}(R) \subset \mathrm{SL}_{2n}(R)$. The induction is quite tedious to carry out and we omit it here.

1.3 Siegel Upper Half-Space

The Siegel upper half-space of genus n is defined by

$$\mathbb{H}_n := \{ Z \in M_n(\mathbb{C}) : Z = {}^t Z, \mathrm{Im}(Z) > 0 \}. \tag{1.4}$$

Let $g = \begin{bmatrix} A & B \\ C & D \end{bmatrix} \in \mathrm{GSp}_{2n}(\mathbb{R})^+ := \{ g \in \mathrm{GSp}_{2n}(\mathbb{R}) : \mu(g) > 0 \}$. For $Z \in \mathbb{H}_n$, we want to define the action

$$g\langle Z \rangle := (AZ + B)(CZ + D)^{-1}. \tag{1.5}$$

Theorem 1.5 *Let $g = \begin{bmatrix} A & B \\ C & D \end{bmatrix} \in \mathrm{GSp}_{2n}(\mathbb{R})^+$ and let $Z \in \mathbb{H}_n$.*

(i) *Define $J(g, Z) := CZ + D$. Then $J(g, Z)$ is invertible and, for $g_1, g_2 \in \mathrm{GSp}_{2n}(\mathbb{R})^+$, we have*

$$J(g_1 g_2, Z) = J(g_1, g_2\langle Z \rangle) J(g_2, Z). \tag{1.6}$$

(ii) *The matrix $g\langle Z \rangle$ is symmetric and we have*

$$\mathrm{Im}\, g\langle Z \rangle = \mu(g) \, {}^t(C\bar{Z} + D)^{-1}(\mathrm{Im}\, Z)(CZ + D)^{-1}. \tag{1.7}$$

(iii) *The map $Z \mapsto g\langle Z \rangle$ is an action of $\mathrm{GSp}_{2n}(\mathbb{R})^+$ on $\mathbb{H}_n$.*

(iv) *If $Z = X + iY \in \mathbb{H}_n$ and $dZ = dX\, dY$ is the Euclidean measure, then*

$$d\,g\langle Z \rangle = \mu(g)^{n(n+1)} |\det(CZ + D)|^{-2n-2} dZ. \tag{1.8}$$

(v) *The element of volume on $\mathbb{H}_n$ given by*

$$d^* Z := \det(Y)^{-(n+1)} dZ,$$

where $dZ = dX dY$ is the Euclidean element of volume, is invariant under all transformations of the group $\mathrm{GSp}_{2n}(\mathbb{R})^+$:

$$d^* g\langle Z \rangle = d^* Z, \text{ for all } g \in \mathrm{GSp}_{2n}(\mathbb{R})^+.$$

Proof If we know that $J(g, Z)$ is invertible, then (1.6) follows by definition (1.5) of $g\langle Z \rangle$. Let us show non-singularity first for $Z = i 1_n$ and arbitrary g. If $J(g, Z)$ is singular, then considering $J(g, Z)^{\mathrm{t}} \overline{J(g, Z)}$, we get that $C\, {}^{\mathrm{t}}C + D\, {}^{\mathrm{t}}D$ is also singular. But this matrix is symmetric and positive semi-definite. Hence, there is a nonzero column vector T such that ${}^{\mathrm{t}}T(C\, {}^{\mathrm{t}}C + D\, {}^{\mathrm{t}}D)T = 0$. This implies that ${}^{\mathrm{t}}T(C\, {}^{\mathrm{t}}C)T = 0$ and ${}^{\mathrm{t}}T(D\, {}^{\mathrm{t}}D)T = 0$, which gives us ${}^{\mathrm{t}}TC = {}^{\mathrm{t}}TD = 0$. But this means that the rank of the matrix (C, D) is less than n, which is impossible since g is non-singular. To get non-singularity for a general Z, first realize $Z = g\langle i 1_n \rangle$ for a suitable g and then use (1.6) for $Z = i 1_n$. This completes proof of part (i).

To get symmetry of $g\langle Z \rangle$, use (1.2) and (1.3) together with the relation

$${}^{\mathrm{t}}(CZ + D) g\langle Z \rangle (CZ + D) = Z\, {}^{\mathrm{t}}CAZ + {}^{\mathrm{t}}DAZ + Z\, {}^{\mathrm{t}}CB + {}^{\mathrm{t}}DB.$$

A similar computation gives (1.7) as well, proving part (ii) of the theorem. Combining parts (i) and (ii), we get part (iii).

To get part (iv), we compute the Jacobian of the change of variable $Z \to g\langle Z \rangle$. Using (1.7) and (1.8), we get part (v) of the theorem. $\square$

Exercise 1.6 *Fill in the details of the proof of Theorem 1.5.*

Exercise 1.7 *Show that, for any $Z \in \mathbb{H}_n$, we have $\det(Z) \neq 0$.*

1.4 Siegel Modular Forms

Let $\Gamma_n := \mathrm{Sp}_{2n}(\mathbb{Z})$.

Definition 1.8 A function $F : \mathbb{H}_n \to \mathbb{C}$ is called a *Siegel modular form* of weight $k \in \mathbb{N}$ and degree n, with respect to Γ_n, if

(i) F is holomorphic,
(ii) F satisfies

$$F((AZ + B)(CZ + D)^{-1}) = \det(CZ + D)^k F(Z), \text{ for all } \begin{bmatrix} A & B \\ C & D \end{bmatrix} \in \Gamma_n,$$

(iii) If $n = 1$, then F is bounded in $Y \geq Y_0$, for any $Y_0 > 0$.

We denote by $M_k(\Gamma_n)$ the $\mathbb{C}$-vector space of Siegel modular forms of weight k and degree n.

Exercise 1.9 *Let $M_k(\Gamma_n)$ be as above.*

(i) If kn is odd, then show that $M_k(\Gamma_n) = 0$.
(ii) Suppose $F \in M_k(\Gamma_n)$. Show that, for all $Z \in \mathbb{H}_n$, F satisfies

$\quad$ *(a) $F(Z + X) = F(Z)$ for all $X = {}^t X \in M_n(\mathbb{Z})$.*
$\quad$ *(b) $F(gZ\,{}^t g) = \det(g)^k F(Z)$ for all $g \in GL_n(\mathbb{Z})$.*
$\quad$ *(c) $F(-Z^{-1}) = \det(Z)^k F(Z)$.*

Any $F \in M_k(\Gamma_n)$ has a *Fourier expansion*

$$F(Z) = \sum_{\substack{T = {}^t T \geq 0 \\ T \text{ half-integral}}} A(T) e^{2\pi i \operatorname{Tr}(TZ)}. \tag{1.9}$$

The translation invariance from part (ii) (a) of Exercise 1.9 and the holomorphy of F give the Fourier expansion, excepting the fact that $T \geq 0$. For $n = 1$, we get $T \geq 0$ from part (iii) of Definition 1.8. For $n > 1$, it follows from the Koecher principle, which is worked out in the next exercise.

Exercise 1.10 *(Koecher principle) Let $F \in M_k(\Gamma_n)$ be such that*

$$F(Z) = \sum_{T = {}^t T \text{ half-integral}} A(T) e^{2\pi i \operatorname{Tr}(TZ)}.$$

(i) Let us denote by $\mathcal{S} := \{T = {}^t T \text{ half-integral}\}$. Define an equivalence relation $\sim$ on $\mathcal{S}$ as follows. For $T_1, T_2 \in \mathcal{S}$, let $T_1 \sim T_2$ if there exists a $g \in SL_n(\mathbb{Z})$ such that $T_1 = {}^t g T_2 g$. Denote by $\{T\}$ the equivalence class of T under $\sim$. Show that

$$F(Z) = \sum_{\mathcal{S}/\sim} A(T) \sum_{T' \in \{T\}} e^{2\pi i \operatorname{Tr}(T'Z)}.$$

$\quad$ *(Hint: Use part (ii) (b) of Exercise 1.9). Conclude that if $A(T) \neq 0$ then, the series $\sum_{T' \in \{T\}} e^{-2\pi \operatorname{Tr}(T')}$ converges absolutely.*
(ii) Suppose T is not positive semi-definite. Show that there is a matrix $g \in SL_n(\mathbb{Z})$ such that the $(1, 1)$ entry of the matrix ${}^t g T g$ is negative. In particular, we can assume without loss of generality, that the matrix entry $T_{11} < 0$.
(iii) Let T be as in part (ii) above. For any positive integer m, consider the matrix

$$g_m = \begin{bmatrix} 1 & m & \\ & 1 & \\ & & 1_{n-2} \end{bmatrix} \in SL_n(\mathbb{Z}).$$

Use the subsequence $\{{}^t g_m T g_m : m \in \mathbb{N}\}$ in $\{T\}$ to show that $\sum_{T' \in \{T\}} e^{-2\pi \operatorname{Tr}(T')}$ diverges. Hence, conclude that $A(T) = 0$.

Definition 1.11 Define an operator Φ on $F \in M_k(\Gamma_n)$ by

$$(\Phi F)(Z') = \lim_{t \to \infty} F(\begin{bmatrix} Z' & 0 \\ 0 & it \end{bmatrix}), \qquad \text{with } Z' \in \mathbb{H}_{n-1}, t \in \mathbb{R}.$$

Theorem 1.12 *The operator Φ gives a well-defined linear map from $M_k(\Gamma_n)$ to the space $M_k(\Gamma_{n-1})$ (with the convention that $M_k(\Gamma_0) = \mathbb{C}$). If F has the Fourier expansion (1.9), then*

$$(\Phi F)(Z') = \sum_{\substack{T_1 = {}^t T_1 \geq 0 \\ T_1 \in M_{n-1}(\mathbb{Z}) \text{ half-integral}}} A(\begin{bmatrix} T_1 & \\ & 0 \end{bmatrix}) e^{2\pi i \operatorname{Tr}(T_1 Z')}, \quad \text{where } Z' \in \mathbb{H}_{n-1}.$$

Proof The Fourier expansion (1.9) implies that F converges uniformly on sets in $\mathbb{H}_n$ with $Y \geq Y_0$ for any $Y_0 > 0$. Hence, we can interchange the limit and the summation of the Fourier expansion, showing that the limit exists. Let $T = \begin{bmatrix} T_1 & * \\ * & t_{nn} \end{bmatrix}$. Note

$$\lim_{t \to \infty} e^{2\pi i \operatorname{Tr}(T \begin{bmatrix} Z' & \\ & it \end{bmatrix})} = \lim_{t \to \infty} e^{-2\pi t t_{nn}} e^{2\pi i \operatorname{Tr}(T_1 Z')} = \begin{cases} 0 & \text{if } t_{nn} > 0; \\ e^{2\pi i \operatorname{Tr}(T_1 Z')} & \text{if } t_{nn} = 0. \end{cases}$$

The shape of the Fourier expansion of ΦF follows from this, together with the observation that, if $t_{nn} = 0$ then the last row and column of T are zero.

To get the automorphy of ΦF with respect to Γ_{n-1} consider the following. Let $g_1 = \begin{bmatrix} A_1 & B_1 \\ C_1 & D_1 \end{bmatrix} \in \Gamma_{n-1}$ and $Z_1 \in \mathbb{H}_{n-1}$. Then, for $Z = \begin{bmatrix} Z_1 & 0 \\ 0 & it \end{bmatrix}$,

$$g = \begin{bmatrix} A_1 & 0 & B_1 & 0 \\ 0 & 1 & 0 & 0 \\ C_1 & 0 & D_1 & 0 \\ 0 & 0 & 0 & 1 \end{bmatrix} \in \Gamma_n, \quad g\langle Z \rangle = \begin{bmatrix} g_1 \langle Z_1 \rangle & 0 \\ 0 & it \end{bmatrix}, \quad \det(J(g, Z)) = \det(J(g_1, Z_1)).$$

$\square$

Definition 1.13 A Siegel modular form $F \in M_k(\Gamma_n)$ is called a *Siegel cusp form* if F lies in the kernel of the Φ operator. Denote the space of cusp forms by $S_k(\Gamma_n)$.

Corollary 1.14 *Let $F \in M_k(\Gamma_n)$. Then $F \in S_k(\Gamma_n)$ if and only if $A(T) = 0$ unless $T > 0$.*

This is proved in Sect. 5, Proposition 2 of [46]. For $F \in S_k(\Gamma_n)$ and $G \in M_k(\Gamma_n)$, define the Petersson inner product by

$$\langle F, G \rangle := \int_{\Gamma_n \backslash \mathbb{H}_n} F(Z) \overline{G(Z)} \det(Y)^k \frac{dX \, dY}{\det(Y)^{n+1}}. \tag{1.10}$$

Chapter 2
Examples

In this chapter, we will present several examples of Siegel modular forms. In addition to theta series and Eisenstein series, we also introduce the Saito–Kurokawa lifts. These are concrete examples of cuspidal Siegel modular forms constructed from elliptic cusp forms. Finally, we consider Siegel modular forms with level $N > 1$ in the genus $n = 2$ case, corresponding to the standard congruence subgroups.

2.1 Examples

2.1.1 Theta Series

Recall notations from Sect. 1.1: let $m, n \in \mathbb{N}$, $A \in M_m(\mathbb{Z})$, A even, $A = {}^t A$, $A > 0$. For $G \in M_{m,n}(Z)$, define the quadratic form $Q(G) = \frac{1}{2} {}^t G A G$. The theta series is defined by

$$\Theta_Q(Z) := \sum_{G \in M_{m,n}(Z)} e^{2\pi i \operatorname{Tr}(Q(G)Z)} = \sum_{\substack{T = {}^t T \geq 0 \\ T \text{ half-integral}}} r_Q(T) e^{2\pi i \operatorname{Tr}(TZ)}, \qquad Z \in \mathbb{H}_n.$$

Here, $r_Q(t)$ is defined as in (1.1). One can show that if $\det A = 1$, then $\Theta_Q \in M_{m/2}(\Gamma_n)$. This is quite complicated and uses the Poisson summation formula. By the way, one can show that there exists $A \in M_m(\mathbb{Z})$, $A > 0$, $A = {}^t A$, $\det A = 1$, A even $\Leftrightarrow 8 | m$. See Theorem III 3.6 of [28] or page 100 of [46] for details.

2.1.2 Eisenstein Series

Let k be a positive integer. Define

© Springer Nature Switzerland AG 2019
A. Pitale, *Siegel Modular Forms*, Lecture Notes in Mathematics 2240,
https://doi.org/10.1007/978-3-030-15675-6_2

$$E_k^{(n)}(Z) := \sum_{\left[\begin{smallmatrix} A & B \\ C & D \end{smallmatrix}\right] \in \Gamma_{0,n} \backslash \Gamma_n} \det(CZ + D)^{-k},$$

where $\Gamma_{0,n} := \{\begin{bmatrix} A & B \\ 0 & D \end{bmatrix} \in \Gamma_n\}$. The convergence of this series is shown in Theorem 1 of Sect. 5 in [46]. The next exercise shows that, if k is even and $k > n + 1$, then $0 \neq E_k^{(n)} \in M_k(\Gamma_n)$.

Exercise 2.1 *Show that* $E_k^{(n)} \in M_k(\Gamma_n)$.

Let $0 \leq r \leq n$ and let $f \in S_k(\Gamma_r)$, with k even. Any $Z \in \mathbb{H}_n$ can be written as $Z = \begin{bmatrix} Z_1 & Z' \\ {}^t Z' & Z_2 \end{bmatrix}$, with $Z_1 \in \mathbb{H}_r$ and $Z_2 \in \mathbb{H}_{n-r}$. Set $Z^* = Z_1 \in \mathbb{H}_r$. Define the series

$$E_{n,r,k}(f)(Z) := \sum_{g = \left[\begin{smallmatrix} A & B \\ C & D \end{smallmatrix}\right] \in P_r \backslash \Gamma_n} f(g\langle Z \rangle^*) \det(CZ + D)^{-k}.$$

Here,

$$P_r := \{ \begin{bmatrix} A' & 0 & B' & * \\ * & U & * & * \\ C' & 0 & D' & * \\ 0 & 0 & 0 & {}^t U^{-1} \end{bmatrix} \in \Gamma_n : \begin{bmatrix} A' & B' \\ C' & D' \end{bmatrix} \in \Gamma_r, U \in \mathrm{GL}_{n-r}(\mathbb{Z}) \}.$$

Note that, if $r = 0$, then $P_0 = \Gamma_{n,0}$ and $E_{n,r,k}(f) = E_k^{(n)}$.

Exercise 2.2 *Show that* $E_{n,r,k}(f)$ *is well defined, i.e.,* $f(g\langle Z \rangle^*) \det(CZ + D)^{-k}$ *is invariant under* P_r.

Theorem 2.3 *Let* $n \geq 1$ *and* $0 \leq r \leq n$ *and* $k > n + r + 1$ *be integers with* k *even. For every cusp form* $f \in S_k(\Gamma_r)$, *the series* $E_{n,r,k}(f)$ *converges to a classical Siegel modular form of weight* k *in* $M_k(\Gamma_n)$ *and* $\Phi^{n-r} E_{n,r,k}(f) = f$.

See page 199–200 of [59] for a proof of the above theorem. This is also proved in Theorem 1 and Proposition 5 of Sect. 5 in [46]. The $E_{n,r,k}(f)$ are called the Klingen–Eisenstein series. We can use this to show that $\Phi : M_k(\Gamma_n) \to M_k(\Gamma_{n-1})$ is surjective for even $k > 2n$ (see corollary to Proposition 5 of Sect. 5 in [46]). There are no good Eisenstein series when k is odd. For example, if $k \equiv n \equiv 1 \pmod 2$, then $M_k(\Gamma_n) = \{0\}$. On the other hand, for sufficiently large k, the space $M_k(\Gamma_{n-1})$ is nontrivial.

Let $\mathcal{E}_k(\Gamma_n)$ be the subspace of $M_k(\Gamma_n)$ generated by the Klingen–Eisenstein series $\{E_{n,r,k}(f) : 0 \leq r \leq n, f \in S_k(\Gamma_r)\}$. If n, k are even and $k > 2n$, then Proposition 7 of Sect. 5 of [46] tells us that

$$M_k(\Gamma_n) = \mathcal{E}_k(\Gamma_n) \oplus S_k(\Gamma_n).$$

Here, the orthogonal direct sum is with respect to the Petersson inner product defined in (1.10).

One can show that there is a $\mu_n > 0$ (depending only on n) such that, whenever the Fourier coefficients of $F \in S_k(\Gamma_n)$ satisfy $A(T) = 0$ for all $T > 0$ with $\mathrm{Tr}(T) < k/\mu_n$, then we have $F \equiv 0$. See page 205 of [77] for this statement and its variants. Also see the theorem on page 642 of [99]. This gives us

$$\dim(S_k(\Gamma_n)) \leq \#\{ \text{half-integral } T = {}^tT > 0 : \mathrm{Tr}(T) < k/\mu_n\} = O(k^N), \; N = \frac{n(n+1)}{2}.$$

Exercise 2.4 *Suppose n, k are even and $k > 2n$, show that $\dim(M_k(\Gamma_n)) \leq O(k^N)$ with $N = \frac{n(n+1)}{2}$.*

Exercise 2.5 *(Cusp forms of genus 2 and weight 10) Normalize the Eisenstein series $E_k^{(n)}$ so that the constant term (coefficient corresponding to $T = 0_n$) is 1, and denote the normalized Eisenstein series by $\tilde{E}_k^{(n)}$. For $n = 1$, we know the structure of the graded ring $M_*(\Gamma_1) = \oplus_k M_k(\Gamma_1)$. It is a polynomial ring generated by the Eisenstein series $e_4 = \tilde{E}_4^{(1)}$ and $e_6 = \tilde{E}_6^{(1)}$.*

(i) *Using $\dim M_{10}(\Gamma_1) = 1$ conclude that $e_{10} - e_4 e_6 = 0$.*
(ii) *Use the Siegel Φ operator to show that $\chi_{10} = \tilde{E}_{10}^{(2)} - \tilde{E}_4^{(2)} \tilde{E}_6^{(2)}$ is a cusp form in $M_{10}(\Gamma_2)$.*

On page 199 of [103], it is shown that χ_{10} is nonzero. The main idea is to write $Z = \begin{bmatrix} \tau_1 & z \\ z & \tau_2 \end{bmatrix}$ and consider the Taylor series of χ_{10} about $z = 0$. One gets $\chi_{10}(Z) = c e^{2\pi i \tau_1} e^{2\pi i \tau_2} z^2 + O(z^3)$ for $c \neq 0$. Hence $\chi_{10} \neq 0$. One can similarly construct nonzero cusp forms $\chi_{12} = \tilde{E}_{12}^{(2)} - \tilde{E}_6^{(2)} \tilde{E}_6^{(2)}$ and χ_{35} of weights 12 and 35, respectively.

Theorem 2.6 (Igusa [42, page 849]) *We have*

$$M_*(\Gamma_2) := \oplus_k M_k(\Gamma_2) = \mathbb{C}[E_4^{(2)}, E_6^{(2)}, \chi_{10}, \chi_{12}, \chi_{35}]/(\chi_{35}^2 = R),$$

where R is a polynomial in $E_4^{(2)}, E_6^{(2)}, \chi_{10}, \chi_{12}$.

2.1.3 Saito–Kurokawa Lifts

Let $f \in S_{2k-2}(\Gamma_1)$ be a cusp form of genus one and weight $2k - 2$, with k even. Kohnen [48, Theorem 1] gave a one-to-one correspondence between the space $S_{2k-2}(\Gamma_1)$ and the space $S_{k-1/2}^+(\Gamma_0(4))$. The latter space is the space of cusp forms of weight $k - 1/2$ with respect to $\Gamma_0(4) = \{ \begin{bmatrix} a & b \\ c & d \end{bmatrix} \in \mathrm{SL}_2(\mathbb{Z}) : 4|c\}$, whose nth Fourier coefficient vanishes whenever $(-1)^{k-1}n \equiv 2, 3 \pmod 4$. Let $g \in S^+_{k-1/2}(\Gamma_0(4))$ correspond to f, and let g have Fourier coefficients $\{c(n)\}$. For a positive definite, symmetric, and half-integral 2×2 matrix T, define

$$A(T) := \sum_{d|\gcd(T)} c\left(\frac{\det(2T)}{d^2}\right) d^{k-1}. \tag{2.1}$$

Theorem 2.7 *Define* $F_f : \mathbb{H}_2 \to \mathbb{C}$ *by the Fourier expansion* $\sum_T A(T) e^{2\pi i \operatorname{Tr}(TZ)}$, *with $A(T)$ given above. Then $F_f \in S_k(\Gamma_2)$ and $F_f \neq 0$ if $f \neq 0$.*

See [60, 61] or page 74 of [25] for the proof using Fourier–Jacobi expansions. In [24], Duke and Imamoglu gave another proof using a converse theorem of Imai [43].

Exercise 2.8 *Write* $A(T) = A(n, r, m)$ *for* $T = \begin{bmatrix} n & r/2 \\ r/2 & m \end{bmatrix}$. *Show that, if F_f is the Saito–Kurokawa lift of f, then the Fourier coefficients satisfy the recurrence relations*

$$A(n, r, m) = \sum_{d|\gcd(n,r,m)} d^{k-1} A\left(\frac{nm}{d^2}, \frac{r}{d}, 1\right). \tag{2.2}$$

These are called the Maass relations. It was shown in [60, 61] that these recurrence relations between Fourier coefficients determine Saito–Kurokawa lifts.

2.2 Congruence Subgroups

Let N be a positive integer. The principal congruence subgroup of level N and genus n is defined by

$$\Gamma_n(N) := \{g \in \Gamma_n : g \equiv 1_n \pmod N\}. \tag{2.3}$$

A subgroup of the symplectic group $\operatorname{Sp}_{2n}(\mathbb{Q})$ is called a *congruence subgroup* if it contains some principal congruence subgroup with finite index.

2.2.1 Congruence Subgroups in Genus 2

We have the following 4 congruence subgroups of $\operatorname{Sp}_4(\mathbb{Q})$.

  (i) Borel congruence subgroup

$$B(N) := \mathrm{Sp}_4(\mathbb{Z}) \cap \begin{bmatrix} \mathbb{Z} & N\mathbb{Z} & \mathbb{Z} & \mathbb{Z} \\ \mathbb{Z} & \mathbb{Z} & \mathbb{Z} & \mathbb{Z} \\ N\mathbb{Z} & N\mathbb{Z} & \mathbb{Z} & \mathbb{Z} \\ N\mathbb{Z} & N\mathbb{Z} & N\mathbb{Z} & \mathbb{Z} \end{bmatrix}.$$

(ii) Siegel congruence subgroup

$$\Gamma_0^{(2)}(N) := \mathrm{Sp}_4(\mathbb{Z}) \cap \begin{bmatrix} \mathbb{Z} & \mathbb{Z} & \mathbb{Z} & \mathbb{Z} \\ \mathbb{Z} & \mathbb{Z} & \mathbb{Z} & \mathbb{Z} \\ N\mathbb{Z} & N\mathbb{Z} & \mathbb{Z} & \mathbb{Z} \\ N\mathbb{Z} & N\mathbb{Z} & \mathbb{Z} & \mathbb{Z} \end{bmatrix}.$$

(iii) Klingen congruence subgroup

$$Q(N) := \mathrm{Sp}_4(\mathbb{Z}) \cap \begin{bmatrix} \mathbb{Z} & N\mathbb{Z} & \mathbb{Z} & \mathbb{Z} \\ \mathbb{Z} & \mathbb{Z} & \mathbb{Z} & \mathbb{Z} \\ \mathbb{Z} & N\mathbb{Z} & \mathbb{Z} & \mathbb{Z} \\ N\mathbb{Z} & N\mathbb{Z} & N\mathbb{Z} & \mathbb{Z} \end{bmatrix}.$$

(iii) Paramodular subgroup

$$K(N) := \mathrm{Sp}_4(\mathbb{Q}) \cap \begin{bmatrix} \mathbb{Z} & N\mathbb{Z} & \mathbb{Z} & \mathbb{Z} \\ \mathbb{Z} & \mathbb{Z} & \mathbb{Z} & N^{-1}\mathbb{Z} \\ \mathbb{Z} & N\mathbb{Z} & \mathbb{Z} & \mathbb{Z} \\ N\mathbb{Z} & N\mathbb{Z} & N\mathbb{Z} & \mathbb{Z} \end{bmatrix}.$$

Exercise 2.9 *Find a matrix $A = \mathrm{diag}(1, 1, t_1, t_2)$, with $t_1, t_2 \in \mathbb{Z}$ and $t_1 | t_2$, such that $K(N) = \mathrm{Sp}_4(\mathbb{Q}) \cap A M_4(\mathbb{Z}) A^{-1}$.*

2.2.2 Siegel Modular Forms with Level

Let k be a positive integer and Γ one of the congruence subgroups above. The space of Siegel modular forms of weight k with respect to Γ is denoted by $M_k(\Gamma)$. Let $f \in S_k(\Gamma_0(N))$ be a newform, where $\Gamma_0(N) = \{ \begin{bmatrix} a & b \\ c & d \end{bmatrix} \in \mathrm{SL}_2(\mathbb{Z}) : N|c \}$. For $Z \in \mathbb{H}_2$, define

$$E_1(Z) := \sum_{g \in (Q(\mathbb{Q}) \cap \Gamma_0^{(2)}(N)) \backslash \Gamma_0^{(2)}(N)} f(g\langle Z \rangle^*) \det(J(g, Z))^{-k},$$

and

$$E_2(Z) := \sum_{g \in D(N) \backslash K(N^2)} f((L_N g)\langle Z \rangle^*) \det(J(L_N g, Z))^{-k},$$

where

$$L_N = \begin{bmatrix} 1 & N & & \\ & 1 & & \\ & & 1 & \\ & & -N & 1 \end{bmatrix}, \text{ and } D_N = L_N^{-1} Q(\mathbb{Q}) L_N \cap K(N^2).$$

Theorem 2.10 (Schmidt, Shukla [94]) *With E_1, E_2 defined above, $E_1 \in M_k(\Gamma_0^{(2)}(N))$ and $E_2 \in M_k(K(N^2))$.*

Brown and Agarwal [1] constructed Saito–Kurokawa liftings starting from elliptic cusp forms $f \in S_{2k-2}(\Gamma_0(N))$, where k is even, and N is odd, square-free, to obtain $F_f \in S_k(\Gamma_0^{(2)}(N))$. If $f \in S_2(\Gamma_0(N_1))$ and $g \in S_{2k}(\Gamma_0(N_2))$ satisfying certain hypothesis, then one can construct a cusp form $F \in S_{k+1}(\Gamma_0^{(2)}(N))$, where $N = \text{lcm}(N_1, N_2)$. This is called the Yoshida lift of f and g, and we will discuss this in more details in Sect. 4.3. One thing to note is that Yoshida lifts do not exist for full level.

Chapter 3
Hecke Theory and L-Functions

In this chapter, we introduce the symplectic Hecke algebra and discuss its action on the Siegel modular forms. This allows us to consider a basis of $M_k(\Gamma_n)$ consisting of simultaneous eigenforms of the Hecke algebra. We explicate the relation between the Hecke eigenvalues and the Fourier coefficients of the modular forms. For genus greater than 1, this relation is very complicated. Finally, we introduce the two L-functions associated with Hecke eigenforms—the spin L-function and the standard L-function.

3.1 The Hecke Algebra

Let $G_n := \mathrm{GSp}_{2n}(\mathbb{Q})^+ \cap M_{2n}(\mathbb{Z})$. Denote by $L(\Gamma_n, G_n)$ the free $\mathbb{Z}$-module consisting of all formal finite linear combinations, with coefficients in $\mathbb{Z}$, of the right cosets $\Gamma_n g$, $g \in \Gamma_n \backslash G_n$. G_n acts on $L(\Gamma_n, G_n)$ by multiplication on the right, and let $\mathcal{H}_n = L(\Gamma_n, G_n)^{\Gamma_n}$ be the space of Γ_n-invariants. Hence, $\mathcal{H}_n$ is the submodule of $L(\Gamma_n, G_n)$ consisting of elements satisfying

$$\sum_i a_i \Gamma_n g_i g = \sum_i a_i \Gamma_n g_i, \text{ for all } g \in \Gamma_n.$$

Exercise 3.1 *For any $g \in G_n$, we have the following finite coset decomposition*

$$\Gamma_n g \Gamma_n = \bigsqcup_i \Gamma_n g_i.$$

Denote $\sum_i \Gamma_n g_i \in L(\Gamma_n, G_n)$ also by $\Gamma_n g \Gamma_n$. Show that $\Gamma_n g \Gamma_n \in \mathcal{H}_n$.

© Springer Nature Switzerland AG 2019
A. Pitale, *Siegel Modular Forms*, Lecture Notes in Mathematics 2240,
https://doi.org/10.1007/978-3-030-15675-6_3

We can define a product on $\mathcal{H}_n$ as follows. If

$$T_1 = \sum_{g \in \Gamma_n \backslash G_n} a_g \Gamma_n g, \qquad T_2 = \sum_{g' \in \Gamma_n \backslash G_n} a_{g'} \Gamma_n g' \in \mathcal{H}_n$$

then set

$$T_1 \cdot T_2 := \sum_{g, g' \in \Gamma_n \backslash G_n} a_g a_{g'} \Gamma_n g g'.$$

Exercise 3.2 *Show that the above product is well defined.*

The space $\mathcal{H}_n$, together with the above multiplication, is called the Hecke algebra. We enumerate certain facts about the structure of $\mathcal{H}_n$. (See Chap. 3 of [4] for details).

(i) $\mathcal{H}_n$ is generated by the double cosets $\Gamma_n g \Gamma_n$, $g \in G_n$.

(ii) Suppose $\Gamma_n g \Gamma_n = \bigsqcup_i \Gamma_n g_i$ and $\Gamma_n g' \Gamma_n = \bigsqcup_i \Gamma_n g'_i$, for some $g, g' \in G_n$. Then

$$\Gamma_n g \Gamma_n \cdot \Gamma_n g' \Gamma_n = \sum_{\Gamma_n h \Gamma_n \subset \Gamma_n g \Gamma_n g' \Gamma_n} c(g, g'; h) \Gamma_n h \Gamma_n,$$

where $c(g, g'; h) = \#\{(i, j) : g_i g'_j \in \Gamma_n h\}$.

(iii) *(symplectic divisors)* Given $g \in G_n$, there is a unique representative in $\Gamma_n g \Gamma_n$ of the form

$$\mathrm{sd}(g) = \mathrm{diag}(d_1, \cdots, d_n, e_1, \cdots e_n),$$

where

$$d_i, e_i \in \mathbb{N}, d_i e_i = \mu(g) \text{ for } i = 1, \cdots, n \text{ and } d_1 | d_2 | \cdots | d_{n-1} | d_n | e_n | e_{n-1} | \cdots | e_1.$$

See Theorem 3.28 of [4] for the above result. It is shown in Theorem 3.31 of [4] that, if $\gcd(e_1/d_1, e'_1/d'_1) = 1$, then

$$\left(\Gamma_n \mathrm{diag}(d_1, \cdots, d_n, e_1, \cdots e_n) \Gamma_n\right) \left(\Gamma_n \mathrm{diag}(d'_1, \cdots, d'_n, e'_1, \cdots e'_n) \Gamma_n\right)$$
$$= \left(\Gamma_n \mathrm{diag}(d_1 d'_1, \cdots, d_n d'_n, e_1 e'_1, \cdots e_n e'_n) \Gamma_n\right).$$

(iv) For a prime number p, let $G_{n,p} := \{g \in G_n : \mu(g) \text{ is a power of } p\}$. Define $\mathcal{H}_{n,p}$ in the same way as $\mathcal{H}_n$ with $G_{n,p}$ instead of G_n. Then, using (iii) above (also see Theorem 3.37 of [4]), we can see that

$$\mathcal{H}_n = \otimes'_p \mathcal{H}_{n,p}.$$

Any element in the restricted tensor product above has the identity of $\mathcal{H}_{n,p}$ as a component for all but finitely many primes p. The algebra $\mathcal{H}_{n,p}$ is called the local Hecke algebra at the prime p.

(v) The local Hecke algebra at the prime p is generated by the $n + 1$ elements

$$T(p) := \Gamma_n \begin{bmatrix} 1_n \\ & p1_n \end{bmatrix} \Gamma_n, \quad T_i(p^2) := \Gamma_n \begin{bmatrix} 1_i \\ & p1_{n-i} \\ & & p^2 1_i \\ & & & p1_{n-i} \end{bmatrix} \Gamma_n, \quad 0 \le i \le n - 1.$$

The elements $T(p)$ and $T_i(p^2)$'s are algebraically independent. Please refer to Theorem 3.40 of [4].

(vi) $\mathcal{H}_n$ is a commutative algebra with identity (Theorem 3.30 of [4]).

(vii) For every positive integer m, define

$$T(m) := \sum_{\mu(g)=m} \Gamma_n g \Gamma_n.$$

Then, for a prime p, $T(p)$ is the same as the one above. On the other hand, $T(p^2)$ is sum of all the $T_i(p^2)$, $0 \le i \le n - 1$ together with $T_n(p^2)$. (This is defined by the same formula as that of $T_i(p^2)$ above allowing $i = n$).

Exercise 3.3 *Consider the formal Dirichlet series*

$$D(s) := \sum_{m=1}^{\infty} T(m) m^{-s}.$$

Show that $D(s)$ has a (formal) expansion as an Euler product of the form

$$D(s) = \prod_{p \text{ prime}} D_p(s), \quad \text{where } D_p(s) = \sum_{r=0}^{\infty} T(p^r) p^{-rs}.$$

3.2 Action of the Hecke Algebra on Siegel Modular Forms

For $g \in G_n$, let $T(g) := \Gamma_n g \Gamma_n = \sqcup_i \Gamma_n g_i \in \mathcal{H}_n$. The Hecke algebra $\mathcal{H}_n$ acts on $M_k(\Gamma_n)$ (or on $S_k(\Gamma_n)$) as follows. Let $F \in M_k(\Gamma_n)$. Then

$$T(g)F := \sum_i F|_k g_i, \tag{3.1}$$

where the slash $|_k$ action is given by

$$\left(F|_k g\right)(Z) := \mu(g)^{nk - \frac{n(n+1)}{2}} \det(CZ + D)^{-k} F(g\langle Z\rangle), \quad \text{for } g = \begin{bmatrix} A & B \\ C & D \end{bmatrix} \in \text{GSp}_{2n}(\mathbb{R})^+. \tag{3.2}$$

Exercise 3.4 *Let the Hecke algebra action be given as in (3.1).*

(i) *Show that this action maps $M_k(\Gamma_n)$ to $M_k(\Gamma_n)$. (In fact, it also maps $S_k(\Gamma_n)$ to $S_k(\Gamma_n)$).*

(ii) *For $F, G \in M_k(\Gamma_n)$, at least one of which is a cusp form, show that*

$$\langle T(g)F, G \rangle = \langle F, T(g)G \rangle, \text{ for all } g.$$

Theorem 3.5 (Andrianov [4, Theorem 4.7]) *The space $M_k(\Gamma_n)$ has a basis of simultaneous eigenfunctions of the Hecke algebra $\mathcal{H}_n$. Furthermore, $S_k(\Gamma_n)$ has such a basis of eigenforms which is orthogonal with respect to the Petersson inner product.*

Let $F \in M_k(\Gamma_n)$ be a Hecke eigenform, and let $T(g)F = \lambda(g)F$, where $\lambda(g)$ are the Hecke eigenvalues. For any prime number p, it is known that there are $n+1$ complex numbers $\alpha_{0,p}, \alpha_{1,p}, \cdots, \alpha_{n,p}$, depending on F, with the following property. If g satisfies $\mu(g) = p^r$, then

$$\lambda(g) = \alpha_{0,p}^r \sum_i \prod_{j=1}^n (\alpha_{j,p} p^{-j})^{d_{ij}}, \tag{3.3}$$

where $\Gamma_n g \Gamma_n = \bigsqcup_i \Gamma_n g_i$, with

$$g_i = \begin{bmatrix} A_i & B_i \\ 0 & D_i \end{bmatrix} \quad \text{and} \quad D_i = \begin{bmatrix} p^{d_{i1}} & & * \\ & \ddots & \\ 0 & & p^{d_{in}} \end{bmatrix}.$$

The $\alpha_{0,p}, \alpha_{1,p}, \cdots, \alpha_{n,p}$ are called the *classical Satake p-parameters* of the eigenform F. Formula (3.3) follows from the Satake map for the p-adic Hecke algebra introduced in Sect. 6.3. For a more classical approach to (3.3), we refer to Formula (1.3.13) of [2].

Let $T(m)F = \lambda(m)F$. From Exercise 3.3, it follows that

$$D_F(s) := \sum_{m=1}^\infty \lambda(m) m^{-s} = \prod_{p \text{ prime}} D_{p,F}(s), \text{ where } D_{p,F}(s) = \sum_{r=0}^\infty \lambda(p^r) p^{-rs}.$$

Theorem 3.6 (Andrianov [2, Theorem 1.3.2]) *For any prime p, we have $D_{p,F}(s) = P(p^{-s})Q(p^{-s})^{-1}$, where*

$$Q(X) = (1 - \alpha_{0,p}X) \prod_{\delta=1}^n \prod_{1 \le i_1 < \cdots < i_\delta \le n} (1 - \alpha_{0,p}\alpha_{i_1,p} \cdots \alpha_{i_\delta,p} X)$$

is a polynomial of degree 2^n and

$$P(X) = \sum_{i=0}^{2^n-2} \phi_i(\alpha_{1,p}, \cdots, \alpha_{n,p})\alpha_{0,p}^i X^i$$

is a polynomial of degree $2^n - 2$ and ϕ_i are some symmetric polynomials in $\alpha_{1,p}, \cdots, \alpha_{n,p}$, with $\phi_1 \equiv 1$ and $\phi_{2^n-2}(\alpha_{1,p}, \cdots, \alpha_{n,p}) = p^{-\frac{(n-1)n}{2}}(\alpha_{1,p}, \cdots, \alpha_{n,p})^{2^{n-1}-1}$.

Exercise 3.7 *(Action of the Hecke operators on Fourier coefficients) Let $n = 2$ and consider the Hecke operator $T(p)$. We have the following coset decomposition*

$$T(p) = \Gamma_2\begin{bmatrix} 1_2 \\ & p1_2 \end{bmatrix}\Gamma_2 = \Gamma_2\begin{bmatrix} p1_2 \\ & 1_2 \end{bmatrix} \sqcup \bigsqcup_{a\in\mathbb{Z}/p\mathbb{Z}} \Gamma_2\begin{bmatrix} 1 & a \\ & p \\ & & p \\ & & & 1 \end{bmatrix}$$

$$\sqcup \bigsqcup_{\alpha,d\in\mathbb{Z}/p\mathbb{Z}} \Gamma_2\begin{bmatrix} p \\ -\alpha & 1 & d \\ & & 1 & \alpha \\ & & & p \end{bmatrix} \sqcup \bigsqcup_{a,b,d\in\mathbb{Z}/p\mathbb{Z}} \Gamma_2\begin{bmatrix} 1 & a & b \\ & 1 & b & d \\ & & p \\ & & & p \end{bmatrix}$$

Suppose $F \in S_k(\Gamma_2)$ with Fourier coefficients $\{A(T)\}$. Applying the Definition 3.1 of the action of $T(p)$ on F, show that the Tth Fourier coefficient of $T(p)F$ is given by

$$A(pT) + p^{2k-3}A(\frac{1}{p}T) + p^{k-2}\Big(\sum_{\alpha\in\mathbb{Z}/p\mathbb{Z}} A(\frac{1}{p}\begin{bmatrix} 1 & \alpha \\ & p \end{bmatrix}T\begin{bmatrix} 1 \\ \alpha & p \end{bmatrix}) + A(\frac{1}{p}\begin{bmatrix} p \\ & 1 \end{bmatrix}T\begin{bmatrix} p \\ & 1 \end{bmatrix})\Big).$$

Here, we assume that $A(S) = 0$, if S is not half-integral.

3.3   Relation Between Fourier Coefficients and Hecke Eigenvalues for Genus 2

Let $F \in M_k(\Gamma_2)$ be a Hecke eigenform. Let us now state a relation between the Fourier coefficients $\{A(T)\}$ of F, and the Hecke eigenvalues, for Siegel modular forms of genus 2. Suppose T is such that $-\det(2T) = -D$ is a fundamental discriminant. Recall that d is a fundamental discriminant if either $d \equiv 1 \pmod 4$ and square-free, or $d = 4m, m \equiv 2, 3 \pmod 4$ and m is square-free. Suppose $\mathbb{Q}(\sqrt{-D})$ has class number $h(D)$, and let $\{T_1, T_2, \cdots, T_{h(D)}\}$ be a set of $\mathrm{SL}_2(\mathbb{Z})$-representatives of primitive, positive definite, integral, binary quadratic forms of discriminant D. For any character χ of the ideal class group of $\mathbb{Q}(\sqrt{-D})$, Theorem 2.4.1 of [2] shows that

$$L(s-k+2, \chi) \sum_{i=1}^{h(D)} \Big(\sum_{n\geq 1} \frac{A(nT_i)}{n^s}\Big) = \Big(\sum_{i=1}^{h(D)} \chi(T_i)A(T_i)\Big)\zeta(2s-2k+4) \sum_{m\geq 1} \frac{\lambda(m)}{m^s}.$$

For the L-function $L(s - k + 2, \chi)$, refer to p. 84 of [2]. In the special case when $\mathbb{Q}(\sqrt{-D})$ has class number one, the above formula reduces to

$$\sum_{n \geq 1} \frac{A(nT)}{n^s} = A(T) \sum_{m \geq 1} \frac{\lambda(m)}{m^s}.$$

From the above two formulas, it is not clear if two Hecke eigenforms with the same Hecke eigenvalues are equal or even related to each other. Using the work of Arthur [5], Ralf Schmidt was recently able to classify Siegel modular forms of genus 2 in 6 different categories. Specializing to the full level case, he is able to prove the following remarkable result.

Theorem 3.8 (Schmidt [93, Theorem 2.6]) *Let k_1, k_2 be two positive integers and let $F_i \in S_{k_i}(\Gamma_2)$ be Hecke eigenforms with Hecke eigenvalues $\{\lambda_i(m) : m \geq 1\}$ for $i = 1, 2$. Suppose, for almost all prime numbers p, we have $p^{(3/2-k_1)r}\lambda_1(p^r) = p^{(3/2-k_2)r}\lambda_2(p^r)$ for all $r \geq 1$. Then $k_1 = k_2$, and F_1 is a scalar multiple of F_2.*

Exercise 3.9 *For $n = 2$, one knows that*

$$Q(X) = 1 - \lambda(p)X + (\lambda(p)^2 - \lambda(p^2) - p^{2k-4})X^2 - \lambda(p)p^{2k-3}X^3 + p^{4k-6}X^4.$$

Use this to show that the Satake p-parameters of F satisfy

$$\alpha_{0,p}^2 \alpha_{1,p} \alpha_{2,p} = p^{2k-3}.$$

For a general n, we have the formula

$$\alpha_{0,p}^2 \alpha_{1,p} \cdots \alpha_{n,p} = p^{kn - \frac{n(n+1)}{2}}. \tag{3.4}$$

3.4 L-Functions

Let $F \in S_k(\Gamma_n)$ be a Hecke eigenfunction. Let $\alpha_{0,p}, \alpha_{1,p}, \cdots, \alpha_{n,p}$ be the Satake p-parameters of F for any prime p. The degree 2^n *spin* L-function of F is defined by

$$L(s, F, \mathrm{spin}) := \prod_{p \text{ prime}} L_p(s, F, \mathrm{spin}),$$

where

$$L_p(s, F, \mathrm{spin})^{-1} := (1 - \alpha_{0,p}p^{-s}) \prod_{\delta=1}^{n} \prod_{1 \leq i_1 < \cdots < i_\delta \leq n} (1 - \alpha_{0,p}\alpha_{i_1,p} \cdots \alpha_{i_\delta,p}p^{-s}).$$

The spin *L*-function converges in some right half-plane. We can also technically define the *L*-function for $F \in M_k(\Gamma_n)$. In [104], Zarkovskaya has shown that the Siegel operator Φ is Hecke equivariant (maps Hecke eigenforms to Hecke eigenforms), and we have the relation $L(s, F, \text{spin}) = L(s, \Phi F, \text{spin})L(s - k + n, \Phi F, \text{spin})$. This tells us that it is enough to consider the spin *L*-functions for cusp forms.

Exercise 3.10 *Let* $F \in S_k(\Gamma_2)$ *be a Hecke eigenform with Hecke eigenvalues* $\{\lambda(m) : m \geq 1\}$. *Show that*

$$\sum_{m \geq 1} \frac{\lambda(m)}{m^s} = \zeta(2s - 2k + 4)^{-1} L(s, F, \text{spin}).$$

Theorem 3.11 (Andrianov [4, p. 167]) *Let* $n = 2$, *and let* $F \in S_k(\Gamma_2)$ *be a Hecke eigenform.*

(i) *For every prime* p,

$$L_p(s, F, \text{spin})^{-1} = 1 - \lambda(p)p^{-s} + (\lambda(p)^2 - \lambda(p^2) - p^{2k-4})p^{-2s}$$
$$- \lambda(p)p^{2k-3}p^{-3s} + p^{4k-6}p^{-4s}.$$

(ii) *The function* $\Lambda(s, F) := (2\pi)^{-2s}\Gamma(s)\Gamma(s - k + 2)L(s, F, \text{spin})$ *has a meromorphic continuation to all of* $\mathbb{C}$, *and satisfies the functional equation*

$$\Lambda(2k - 2 - s, F) = (-1)^k \Lambda(s, F).$$

(iii) $\Lambda(s, F)$ *has at most two simple poles at* $s = k - 2, k$. *If* k *is odd, then* $\Lambda(s, F)$ *is entire.*

Exercise 3.12 *Set* $\alpha_p = p^{3/2-k}\alpha_{0,p}$ *and* $\beta_p = \alpha_p \alpha_{1,p}$. *Show that*

$$\lambda(p) = p^{k-3/2}(\alpha_p + \alpha_p^{-1} + \beta_p + \beta_p^{-1}),$$
$$\lambda(p^2) = p^{2k-3}\left((\alpha_p + \alpha_p^{-1})^2 + (\alpha_p + \alpha_p^{-1})(\beta_p + \beta_p^{-1}) + (\beta_p + \beta_p^{-1})^2 - 2 - 1/p\right).$$

For $n = 3$, Asgari and Schmidt [6, Theorem 5] proved that the spin *L*-function has a meromorphic continuation. In [76], Pollack proved the functional equation and finiteness of poles. For general n, the meromorphic continuation and functional equation are open conjectures.

Once again, let $F \in S_k(\Gamma_n)$ be a Hecke eigenfunction. Let $\alpha_{0,p}, \alpha_{1,p}, \ldots, \alpha_{n,p}$ be the Satake p-parameters of F for any prime p. The degree $2n + 1$ *standard L*-function of F is defined by

$$L(s, F, \text{std}) = \prod_{p \text{ prime}} L_p(s, F, \text{std}), \tag{3.5}$$

where

$$L_p(s, F, \text{std})^{-1} := (1 - p^{-s}) \prod_{i=1}^{n} (1 - \alpha_{i,p} p^{-s})(1 - \alpha_{i,p}^{-1} p^{-s}).$$

For all n, it is known that $L(s, F, \text{std})$ has a meromorphic continuation to $\mathbb{C}$ with finitely many poles and has a functional equation under $s \mapsto 1 - s$ (see [9]). Once again we can define the standard L-function for non-cusp forms. We have the following relation between the standard L-functions of F and ΦF.

$$L(s, F, \text{std}) = L(s, \Phi F, \text{std})\zeta(s - k + n)\zeta(s + k - n). \tag{3.6}$$

3.4.1 L-Function for Saito–Kurokawa Lifts

Let k be even and let $f \in S_{2k-2}(\Gamma_1)$ be a Hecke eigenform with Fourier coefficients $\{a(n)\}$ normalized by setting $a(1) = 1$. Let $F_f \in S_k(\Gamma_2)$ be the Saito–Kurokawa lift introduced in Sect. 2.1.3. Then F_f is also a Hecke eigenform. One can explicitly compute the Hecke eigenvalues of F_f in terms of the Hecke eigenvalues of f using the definition in (2.1), and the action of the Hecke operators on Fourier coefficients. This computation leads to the following relation.

$$L(s, F_f, \text{spin}) = \zeta(s - k + 1)\zeta(s - k + 2)L(s, f), \quad \text{where } L(s, f) = \sum_{n=1}^{\infty} \frac{a(n)}{n^s}.$$
$$\tag{3.7}$$

For further details on this, we refer the reader to Theorem 6.3 and Corollary 1 on pages 77, 80 of [25]. Note that, $L(s, F_f, \text{spin})$ has a pole at $s = k$. In fact, it was shown by Evdokimov [26, Theorem 1] and Oda [65, p. 324] that $F \in S_k(\Gamma_2)$ is a Saito–Kurokawa lift if and only if $L(s, F_f, \text{spin})$ has a pole at $s = k$.

Exercise 3.13 *We know that* $\xi(s) := \pi^{-s/2}\Gamma(s/2)\zeta(s)$ *satisfies the functional equation* $\xi(s) = \xi(1 - s)$. *Also, if* $g \in S_{k'}(\Gamma_1)$, *then* $\Lambda(s, g) := (2\pi)^{-s}\Gamma(s)L(s, g)$ *satisfies the functional equation* $\Lambda(k' - s, g) = (-1)^{k'/2}\Lambda(s, g)$. *Use these functional equations to derive the functional equation for* $\Lambda(s, F_f)$, *where* F_f *is the Saito–Kurokawa lift of* $f \in S_{2k-2}(\Gamma_1)$.

Exercise 3.14 *Show that*

$$L(s, F_f, \text{std}) = \zeta(s)L(s + k - 1, f)L(s + k - 2, f).$$

Chapter 4
Nonvanishing of Fourier Coefficients and Applications

In this chapter, we start by stating the Generalized Ramanujan conjecture (GRC) for Siegel modular forms. We will discuss progress toward GRC, and its application. Next, we will consider the question of nonvanishing of Fourier coefficients. In the previous chapters, we have seen that if the first few Fourier coefficients of a Siegel modular form F are zero, then F is identically zero. For certain applications, it is useful to know that a nonzero F has nonzero Fourier coefficients corresponding to some special kind of symmetric half-integral matrices. In this chapter, we will explore the results regarding nonvanishing of Fourier coefficients $A(T)$ for *primitive* T and *fundamental* T. We will end with an amazing application to simultaneous nonvanishing of pairs of L-functions and Böcherer's conjecture.

4.1 Generalized Ramanujan Conjecture

Let $F \in S_k(\Gamma_n)$ be a Hecke eigenform with Satake p-parameters $\alpha_{0,p}, \alpha_{1,p}, \ldots, \alpha_{n,p}$.

Conjecture 4.1 (Generalized Ramanujan Conjecture (GRC)) *For all prime numbers p, the Satake p-parameters satisfy*

$$|\alpha_{i,p}| = 1, \text{ for all } i = 1, 2, \cdots, n.$$

The article by Sarnak [91] gives a very good account of GRC. Using (3.4), the GRC will imply that $|\alpha_{0,p}| = p^{kn/2-n(n+1)/4}$. For $n = 1$, the GRC is a theorem for $f \in S_k(\Gamma_0(N), \chi')$ due to Deligne [22]. For $n = 2$, we have counterexamples— namely, the Saito–Kurokawa lifts. Let k be even and let $f \in S_{2k-2}(\Gamma_1)$ be a Hecke eigenform with Satake p-parameters $\beta_{0,p}, \beta_{1,p}$. We know that $|\beta_{0,p}| = p^{k-3/2}$ and $|\beta_{1,p}| = 1$. From (3.7), we can conclude that the Satake p-parameters of the Saito–Kurokawa lift $F_f \in S_k(\Gamma_2)$ satisfy

© Springer Nature Switzerland AG 2019
A. Pitale, *Siegel Modular Forms*, Lecture Notes in Mathematics 2240,
https://doi.org/10.1007/978-3-030-15675-6_4

$$\{\alpha_{0,p}, \alpha_{0,p}\alpha_{1,p}, \alpha_{0,p}\alpha_{2,p}, \alpha_{0,p}\alpha_{1,p}\alpha_{2,p}\} = \{p^{k-1}, p^{k-2}, \beta_{0,p}, \beta_{0,p}\beta_{1,p}\}.$$

Hence, $\alpha_{1,p}, \alpha_{2,p}$ have absolute values $p^{1/2}, p^{-1/2}$, violating the GRC.

Theorem 4.2 (Weissauer [106]) *Let $F \in S_k(\Gamma_2)$ be a Hecke eigenform, which is not a Saito–Kurokawa lift. Then, F satisfies the generalized Ramanujan conjecture.*

For a general n, it was shown in [67] that a Hecke eigenform $F \in S_k(\Gamma_n)$ satisfies GRC if and only if for every $\epsilon > 0$, there is a constant $C_\epsilon > 0$, depending only on ϵ, n, p, such that

$$|\lambda(p^r)| \leq C_\epsilon p^{r(\frac{nk}{2} - \frac{n(n+1)}{4})}, \quad \text{for all } r \geq 0.$$

For $n = 2$, the validity of GRC has the following application.

Theorem 4.3 (Farmer, Pitale, Ryan, Schmidt [27, Theorem 1.3]) *Let k_j be positive integers for $j = 1, 2$. Suppose $F_j \in S_{k_j}(\Gamma_2)$ are Siegel Hecke eigenforms with Hecke eigenvalues $\lambda_j(n), n \geq 1$. If $p^{3/2-k_1}\lambda_1(p) = p^{3/2-k_2}\lambda_2(p)$ for all but finitely many p, then $k_1 = k_2$ and F_1, F_2 have the same eigenvalues for the Hecke operator $T(n)$ for all n.*

The above theorem follows from a *strong multiplicity one* theorem for Dirichlet series. Using analytic number theoretic techniques, one can show that if two Dirichlet series $L_1(s) = \sum a_1(n)n^{-s}$ and $L_2(s) = \sum a_2(n)n^{-s}$, satisfying certain suitable hypotheses, have the property that $a_1(p) = a_2(p)$ for almost all p, then $L_1(s) = L_2(s)$.

The remarkable fact here is that the Hecke operator $T(p)$ alone does not generate the local Hecke algebra at p. This Hecke algebra is generated by $T(p)$ and $T(p^2)$. The fact that the coincidence of the eigenvalues for $T(p)$ is enough is, of course, a global phenomenon using GRC.

Exercise 4.4 *Suppose $F_j \in S_{k_j}(\Gamma_2)$ are Siegel Hecke eigenforms that satisfy, for all but finitely many p, $p^{3/2-k_1}\lambda_1(p) = p^{3/2-k_2}\lambda_2(p)$. Show that $k_1 = k_2$ and $F_1 \in \mathbb{C}F_2$.*

4.2 Nonvanishing of Fourier Coefficients

Let us denote by $\mathcal{S}_n$ the set of all $n \times n$ symmetric, half-integral, positive definite matrices.

Problem: *What subsets of $\mathcal{S}_n$ have the property that, if the Fourier coefficient of a Siegel modular form vanishes for all elements in the subset, then the Siegel modular form has to be zero?*

For $T \in \mathcal{S}_n$, the *content* of T is defined by

$$c(T) := \max\{a \in \mathbb{N} : a^{-1}T \text{ is half-integral }\}.$$

We say that T is *primitive* if $c(T) = 1$.

Theorem 4.5 (Zagier [109, pp. 387]) *Let $F \in M_k(\Gamma_2)$ have Fourier coefficients $A(T)$. If $A(T) = 0$ for all primitive matrices T, then $F = 0$.*

The main steps of the proof of the above theorem are listed in the following exercise.

Exercise 4.6 *Let $n = 2$. Write $Z \in \mathbb{H}_2$ as $Z = \begin{bmatrix} \tau & z \\ z & \tau' \end{bmatrix}$, with $\tau, \tau' \in \mathbb{H}_1, z \in \mathbb{C}$, and $T > 0$ as $T = \begin{bmatrix} n & r/2 \\ r/2 & m \end{bmatrix}$ with $n, m, r \in \mathbb{Z}$. Write $F \in S_k(\Gamma_2)$ as*

$$F(\tau, z, \tau') = \sum_{\substack{n,m,r \in \mathbb{Z} \\ n>0, 4mn-r^2>0}} A(n, r, m) e^{2\pi i n \tau} e^{2\pi i m \tau'} e^{2\pi i r z}.$$

(i) *Let $\begin{bmatrix} a & b \\ c & d \end{bmatrix} \in \Gamma_1$ and $s \in \mathbb{Z}$. Using $\begin{bmatrix} a & & b & \\ & 1 & & \\ c & & d & \\ & & & 1 \end{bmatrix}$ and $\begin{bmatrix} 1 & & & \\ s & 1 & & \\ & & 1 & -s \\ & & & 1 \end{bmatrix}$, and the auto-morphy of F, show that*

$$F\left(\frac{a\tau+b}{c\tau+d}, \frac{z}{c\tau+d}, \tau' - \frac{cz^2}{c\tau+d}\right) = (c\tau+d)^k F(\tau, z, \tau'),$$
$$F(\tau, z+s\tau, \tau'+2sz+s^2\tau) = F(\tau, z, \tau').$$

(ii) *For $m > 0$, define the functions $\phi_m(\tau, z)$ on $\mathbb{H}_1 \times \mathbb{C}$ by*

$$F(\tau, z, \tau') = \sum_{m=1}^{\infty} \phi_m(\tau, z) e^{2\pi i m \tau'}. \tag{4.1}$$

Show that the ϕ_m satisfy the following relations.

$$\phi_m\left(\frac{a\tau+b}{c\tau+d}, \frac{z}{c\tau+d}\right) = (c\tau+d)^k e^{\frac{2\pi i m c z^2}{c\tau+d}} \phi_m(\tau, z), \qquad \begin{bmatrix} a & b \\ c & d \end{bmatrix} \in \Gamma_1,$$
$$\phi_m(\tau, z+s\tau) = e^{-2\pi i m (2sz+s^2\tau)} \phi_m(\tau, z), \qquad s \in \mathbb{Z}.$$

Holomorphic functions on $\mathbb{H}_1 \times \mathbb{C}$ satisfying the above two properties are called Jacobi forms of weight k and index m and (4.1) is called the Fourier–Jacobi expansion of F.

(iii) *Take any ϕ_m as above and consider its Taylor expansion about $z = 0$ given by*

$$\phi_m(\tau, z) = \sum_{\nu=0}^{\infty} \lambda_\nu(\tau, z), \qquad \lambda_\nu(\tau, z) = \frac{1}{\nu!} \left(\frac{\partial^\nu}{\partial z^\nu} \phi_m(\tau, z)\right)\big|_{z=0} z^\nu.$$

Find a formula for $\lambda_\nu(\tau, z)$ in terms of the Fourier coefficients $A(n, r, m)$.

(iv) Let ν_0 be the smallest ν such that $\lambda_\nu(\tau, z)$ is not identically zero. Using (ii), show that, for any z, we have $\lambda_{\nu_0}(\tau, z) \in M_{k+\nu_0}(\Gamma_1)$.

(v) Let ℓ be a positive integer and let $f \in M_{k'}(\Gamma_1)$ with Fourier coefficients $\{a(t) : t \geq 0\}$. It is known that if $a(t) = 0$ for all t coprime to ℓ, then f is the zero function. Use this fact and the above steps to complete the proof of Zagier's theorem.

Zagier's result was extended by Yamana to higher genus and included higher level. For a positive integer N, the Siegel congruence subgroup of level N is given by

$$\Gamma_0^n(N) := \{g = \begin{bmatrix} A & B \\ C & D \end{bmatrix} \in \Gamma_n : C \equiv 0 \pmod{N}\}.$$

Theorem 4.7 (Yamana [107, Theorem 2]) *Let $F \in M_k(\Gamma_0^n(N))$ with Fourier coefficients $A(T)$. Suppose that $A(T) = 0$ for all T such that $c(T)$ divides N. Then $F = 0$.*

If we want to remove the condition on the content, then we have to consider the subspace of new forms. Let $S_k^O(\Gamma_0^n(N))$ be the linear subspace of $S_k(\Gamma_0^n(N))$ spanned by the set

$$\{F(dZ) : F \in S_k(\Gamma_0^n(M)), dM|N, M \neq N\}.$$

Theorem 4.8 (Ibukiyama, Katsurada [39]) *Let $F \in S_k(\Gamma_0^n(N))$ with Fourier coefficients $A(T)$. Assume that F belongs to the orthogonal complement of $S_k^O(\Gamma_0^n(N))$. If $A(T) = 0$ for all primitive matrices T, then $F = 0$.*

In genus 2, from the point of view of representation theory, a smaller subset of fundamental matrices T plays an important role. The discriminant of $T \in S_2$ is defined by $\text{disc}(T) = -\det(2T)$. The matrix T in S_2 is called *fundamental*, if $D = \text{disc}(T)$ is a fundamental discriminant, i.e., either $D \equiv 1 \pmod 4$ and square-free, or $D = 4m$, $m \equiv 2, 3 \pmod 4$ and m is square-free. Observe that if T is fundamental, then it is automatically primitive. Observe also that if D is odd, then T is fundamental if and only if D is square-free.

Theorem 4.9 (Saha [87, Theorem 3.4]) *Let $k > 2$ and N be a square-free positive integer. If $N > 1$, then assume that k is even. Let $0 \neq F \in S_k(\Gamma_0^2(N))$ lie in the orthogonal complement of $S_k^O(\Gamma_0^2(N))$. Then, there are infinitely many fundamental matrices T such that the Fourier coefficients $A(T) \neq 0$.*

Sketch of proof: By Theorem 4.8, there is a primitive matrix T' such that $A(T') \neq 0$. Let $T' = \begin{bmatrix} a & b/2 \\ b/2 & c \end{bmatrix}$. It is a classical result that the primitive quadratic form $ax^2 + bxy + cy^2$ represents infinitely many primes. So, let x_0, y_0 be such that

$ax_0^2 + bx_0y_0 + cy_0^2$ is an odd prime p not dividing N. Since $\gcd(x_0, y_0) = 1$, we can find integers x_1, y_1 such that $A = \begin{bmatrix} y_1 & y_0 \\ x_1 & x_0 \end{bmatrix} \in SL_2(\mathbb{Z})$. Then , $T = {}^tAT'A$ has the property that $A(T) \neq 0$, and T is of the form $\begin{bmatrix} a_0 & b_0/2 \\ b_0/2 & p \end{bmatrix}$. For all integers n, r with $4np > r^2$, denote by $c(n, r) := A(\begin{bmatrix} n & r/2 \\ r/2 & p \end{bmatrix})$. Now let

$$h(\tau) = \sum_{m=1}^{\infty} c(m)e^{2\pi i m\tau}, \tau \in \mathbb{H}_1, c(m) := \sum_{\substack{0 \leq \mu \leq 2p-1 \\ \mu^2 \equiv -m \pmod{4p}}} c((m + \mu^2)/(4p), \mu).$$

It turns out that $h \in S_{k-1/2}(4pN)$. The modular form h is nonzero. To see that, let $d_0 = 4a_0p - b_0^2$. Then, $c(d_0)$ is equal to $A(\begin{bmatrix} a_0 & b_0/2 \\ b_0/2 & p \end{bmatrix}) + A(\begin{bmatrix} a_0 + p - b_0 & p - b_0/2 \\ p - b_0/2 & p \end{bmatrix})$

$= 2A(\begin{bmatrix} a_0 & b_0/2 \\ b_0/2 & p \end{bmatrix}) \neq 0$. (Conjugate the second matrix by $\begin{bmatrix} 1 & -1 \\ & 1 \end{bmatrix}$). Now, using techniques from analytic number theory, Saha has proved that such half-integral weight modular forms h have infinitely many Fourier coefficients $c(d) \neq 0$, where d is a fundamental discriminant. For any such d, there exists a μ such that $c(\frac{d+\mu^2}{4p}, \mu) = A(\begin{bmatrix} \frac{d+\mu^2}{4p} & \mu/2 \\ \mu/2 & p \end{bmatrix}) \neq 0$.   □

Exercise 4.10 *There is an open conjecture that any $F \in S_k(\Gamma_2)$ has a nonzero Fourier coefficient $A(T)$ with T of the form $\begin{bmatrix} a & b/2 \\ b/2 & 1 \end{bmatrix}$. Show that this is true if F is a Saito–Kurokawa lift.*

4.3 Application of the Nonvanishing Result

One of the main applications of Theorem 4.9 is toward the existence of nontrivial global Bessel models for the automorphic representations corresponding to Siegel cusp forms of genus 2. We will discuss that in detail in a future chapter (see Theorem 8.6). Let us now present an application toward simultaneous nonvanishing of $GL(2) \times GL(2)$ L-functions. Let us fix two elliptic cusp forms f and g. Let K be an imaginary quadratic field of discriminant $-d < 0$, let O_K be its ring of integers, and let Λ be a character of the ideal class group Cl_K of K. Define the function

$$\theta_\Lambda(z) := \sum_{\mathfrak{a}} \Lambda(\mathfrak{a})e^{2\pi i N(\mathfrak{a})z}, \quad z \in \mathbb{H}_1.$$

Here, $\mathfrak{a}$ runs through integral ideals in O_K. Then $\theta_\Lambda \in M_1(\Gamma_0(d), \left(\frac{-d}{*}\right))$, and it is a cusp form if and only if $\Lambda^2 \neq 1$ (see pp. 215 of [30]).

Problem: *Fix elliptic cusp forms f and g. Find K, Λ so that*

$$L(\frac{1}{2}, f \times \theta_\Lambda)L(\frac{1}{2}, g \times \theta_\Lambda) \neq 0.$$

For general f and g, this is a very hard and open problem. Let us now consider a special case. Let N_1, N_2 be square-free positive integers that are not coprime, and let $N = \text{lcm}(N_1, N_2)$. Let $f \in S_2(\Gamma_0(N_1))$ and $g \in S_{2k}(\Gamma_0(N_2))$ be newforms. For every prime p dividing $\gcd(N_1, N_2)$, assume that f and g have the same Atkin–Lehner eigenvalues at p. Then, there exists a nonzero $F \in S_{k+1}(\Gamma_0^2(N))$ with the following properties.

  (i) F is orthogonal to $S_{k+1}^O(\Gamma_0^2(N))$.
 (ii) F is an eigenfunction for the local Hecke algebra at all primes.
(iii) We have
$$L(s, F \times \theta_\Lambda) = L(s, f \times \theta_\Lambda)L(s, g \times \theta_\Lambda).$$

This was first studied by Yoshida and hence is called the Yoshida lift of f and g. See [8, 13, 108] for details on the Yoshida lift. The L-functions on the right-hand side above are Rankin–Selberg L-functions of two elliptic modular forms (see Section 1.6 of [17]). The L-function on the left-hand side is the degree 8 Rankin–Selberg L-function of F and θ_Λ (see [30]).

Note that F satisfies all the hypotheses of Theorem 4.9, and so we can conclude that there are infinitely many fundamental T such that $A(T) \neq 0$. In fact, Saha and Schmidt [89] can prove that there are infinitely many pairs (K, Λ), where K is an imaginary quadratic field and Λ is an ideal class character, such that

$$R(F, K, \Lambda) := \sum_{c \in \text{Cl}_K} \Lambda^{-1}(c)A(c) \neq 0. \tag{4.2}$$

It is well known that the SL$(2, \mathbb{Z})$-equivalence classes of binary quadratic forms of discriminant $-d$ are in natural bijectives correspondence with the elements of Cl$_K$. Since $A({}^t gTg) = A(T)$ for all $g \in \text{SL}(2, \mathbb{Z})$, it follows that the notation $A(c)$ makes sense for $c \in \text{Cl}_K$. Prasad and Takloo-Bighash [78] proved that, for a Yoshida lift F

$$R(F, K, \Lambda) \neq 0 \Rightarrow L(1/2, F \times \theta_\Lambda) \neq 0.$$

For a general F, the above implication is still a conjecture, although Furusawa and Morimoto have solved the problem for a large class in [32]. Now, part (iii) of the properties of Yoshida lifts immediately tells us that there are infinitely many pairs (K, Λ) such that

$$L(\frac{1}{2}, f \times \theta_\Lambda)L(\frac{1}{2}, g \times \theta_\Lambda) \neq 0.$$

4.4 Böcherer's Conjecture

Let $F \in S_k(\Gamma_2)$ with Fourier coefficients $\{A(T) : T > 0\}$. We define

$$R(F, K) := \sum_{c \in \mathrm{Cl}_K} A(c). \tag{4.3}$$

For odd k, it is easy to see that $R(F, K)$ equals 0. If k is even, Böcherer [10] made a remarkable conjecture that relates the quantity $R(F, K)$ to the central value of a certain L-function.

Conjecture 4.11 (Böcherer [10]) *Let k be even and $F \in S_k(\Gamma_2)$ be a nonzero Hecke eigenform. Then there exists a constant c_F, depending only on F, such that for any imaginary quadratic field $K = \mathbb{Q}(\sqrt{d})$, with $d < 0$ a fundamental discriminant, we have*

$$|R(F, K)|^2 = c_F \cdot w(K)^2 \cdot |d|^{k-1} \cdot L(1/2, F \otimes \chi_d).$$

Above, $\chi_d = \left(\frac{d}{\cdot}\right)$ is the Kronecker symbol (i.e., the quadratic Hecke character associated via class field theory to the field $\mathbb{Q}(\sqrt{d})$), $w(K)$ denotes the number of distinct roots of unity inside K, and $L(s, F \otimes \chi_d)$ denotes the spin L-function of F twisted by the character χ_d.

In [23], we computed a conjectural value of the constant c_F as

$$c_F = \frac{2^{4k-4} \cdot \pi^{2k+1}}{(2k-2)!} \cdot \frac{L(1/2, F)}{L(1, F, \mathrm{Ad})} \cdot \langle F, F \rangle. \tag{4.4}$$

Here, $L(s, F, \mathrm{Ad})$ is the adjoint L-function of F. Recently, in a remarkable paper [33], Furusawa and Morimoto have proved a version of Conjecture 4.11 under some hypotheses. In particular, their work implies that Conjecture 4.11 is now a theorem for $k > 2$, with the precise value of c_F given by (4.4).

Chapter 5
Applications of Properties of L-Functions

In this chapter, we present two applications of properties of L-functions of Siegel modular forms. The first one is to determine whether a given modular form is a cusp form based on the size of its Fourier coefficients. The main tool is the domain of convergence of the standard L-function associated to the modular form. The second application is the infinitely many sign changes of Hecke eigenvalues of Siegel–Hecke eigenforms using the properties of the spin L-function.

5.1 Determining Cusp Forms by Size of Fourier Coefficients

Let $F \in M_k(\Gamma_n)$ have Fourier coefficients $A(T)$. Then it is known that

$$|A(T)| \ll_F \det(T)^k.$$

(See Page 143 of [46])

The implied constant depends only on F. In addition, if F is a cusp form in $S_k(\Gamma_n)$, then we have the "Hecke bound" (see Page 170 of [11])

$$|A(T)| \ll_F \det(T)^{k/2}. \tag{5.1}$$

Problem: *Suppose that $F \in M_k(\Gamma_n)$ is such that all its Fourier coefficients $A(T)$, for $T > 0$, satisfy (5.1). Is F automatically a cusp form?*

This was first addressed by Kohnen for elliptic modular forms of weight 2 and level N (see [50]). Degree 2, level 1 case was also addressed by Kohnen and Martin [51]. Both of these use explicit information on the Fourier coefficients of Eisenstein

© Springer Nature Switzerland AG 2019
A. Pitale, *Siegel Modular Forms*, Lecture Notes in Mathematics 2240,
https://doi.org/10.1007/978-3-030-15675-6_5

series that are only available for low genus. In the higher genus case, Böcherer and Das came up with a completely different method to solve the problem.

Theorem 5.1 (Böcherer, Das [11, Theorem 4.1]) *Let $k > 2n$ and let $F \in M_k(\Gamma_n)$ be such that all its Fourier coefficients $A(T)$, for $T > 0$, satisfy (5.1). Then $F \in S_k(\Gamma_n)$.*

They prove this theorem in much more generality to include level N, and allow less restrictions on k. We are looking at the simplest case here. Their proof involves the standard L-function and for that, we need a Hecke eigenform. That is achieved in the following exercise.

Exercise 5.2 *Let $F \in M_k(\Gamma_n)$ be such that all its Fourier coefficients $A(T)$, for $T > 0$, satisfy (5.1). Then, there is a Hecke eigenform $G \in M_k(\Gamma_n)$ such that all its Fourier coefficients $A(T)$, for $T > 0$, satisfy (5.1). In addition, if F is non-cuspidal, then we can choose G to be non-cuspidal.*

Let us now assume that F is a Hecke eigenform. We have already defined in (3.5) the standard L-function $L(s, F, \text{std}) = \prod_p L_p(s, F, \text{std})$ for F. Andrianov has a Dirichlet series formula relating the Fourier coefficients to the standard L-function. Let $T_0 > 0$ be such that $A(T_0) \neq 0$ (this is guaranteed by the condition $k > 2n$) and put $M = \det(2T_0)$. On page 147 of [3], it is shown that

$$\sum_X A({}^tX T_0 X) \det(X)^{-s-k+1} = A(T_0) \Lambda^M(s) L^M(s, F, \text{std}), \qquad (5.2)$$

where

$$\Lambda^M(s) = \begin{cases} L^M(s+m, \chi_{T_0})^{-1} \prod_{i=0}^{m-1} \zeta^M(2s-2i)^{-1} & \text{if } n = 2m; \\ \prod_{i=0}^{m} \zeta^M(2s+2i)^{-1} & \text{if } n = 2m+1. \end{cases}$$

Here, X runs over all nonsingular integral matrices of size n with $(\det X, M) = 1$, modulo the action of $\text{GL}_n(\mathbb{Z})$ from the right. $L(s, \chi_{T_0})$ is the Dirichlet L-series attached to the quadratic character $\left(\frac{(-1)^m \det(2T_0)}{*} \right)$. The superscript M on the L-functions means that we are taking the product over all primes $p \nmid M$.

Condition (5.1) implies that the left-hand side of (5.2) converges absolutely in the region $\text{Re}(s) > n + 1$. Hence, the standard L-function is a nonzero holomorphic function on $\text{Re}(s) > n + 1$. We will get a contradiction to this holomorphy if F is not cuspidal.

So, let us assume that F is not cuspidal. Hence $\Phi F \neq 0$, where Φ is the Siegel operator from Definition 1.11. Recall the relation (3.6) between the standard L-functions of F and ΦF.

$$L(s, F, \text{std}) = L(s, \Phi F, \text{std}) \zeta(s - k + n) \zeta(s + k - n).$$

Exercise 5.3 *Show that we cannot have* $\Phi^n F \neq 0$.

Thus, we can assume that there is a r satisfying $1 \leq r \leq n-1$ such that $\Phi^{n-r} F$ is a nonzero cusp form. This satisfies

$$L(s, F, \text{std}) = L(s, \Phi^{n-r} F, \text{std}) \prod_{i=0}^{n-r-1} \zeta(s - k + n - i)\zeta(s + k - n + i).$$

Since $\Phi^{n-r} F$ is a cusp form, a result of Shimura [96, Theorem A] says that the L-function $L(s, \Phi^{n-r} F, \text{std})$ is a nonzero holomorphic function for $\text{Re}(s) > r/2 + 1$. The rightmost pole of the product of zeta functions is at $s = k - r$. This is not canceled by $L(s, \Phi^{n-r} F, \text{std})$ since it is nonzero at this point. Hence, $s = k - r$ is a pole for $L(s, F, \text{std})$, which again contradicts the holomorphy of the L-function for $\text{Re}(s) > n + 1$. $\qquad\square$

5.2 Sign Changes of Hecke Eigenvalues

Let us now discuss the sign changes of Hecke eigenvalues of Siegel–Hecke eigenforms. For that, we first need to know that the Hecke eigenvalues are real.

Exercise 5.4 *Show that the Hecke eigenvalues of a Hecke eigenform $F \in S_k(\Gamma_n)$ are real.*

Let $F \in S_k(\Gamma_n)$ be a Hecke eigenform with eigenvalues $\lambda(m)$ for $m \geq 1$. We want to explore the question of sign changes in the sequence $\{\lambda(m)\}$.

Theorem 5.5 (Breulmann [14]) *Let $F \in S_k(\Gamma_2)$ be a Hecke eigenform with eigenvalues $\lambda(m)$ for $m \geq 1$. Then F is a Saito–Kurokawa lift if and only if $\lambda(m) > 0$ for all m.*

It was shown by Evdokimov [26] and Oda [65] that F is a Saito–Kurokawa lift if and only if $L(s, F, \text{spin})$ has a pole at $s = k$. Suppose all the Hecke eigenvalues are positive, then, using this criteria, and Landau's theorem [55, Page 536], and the relation between $L(s, F, \text{spin})$ and $\sum \lambda(m)m^{-s}$, one can conclude that F has to be a Saito–kurokawa lift. On the other hand, if F is a Saito–Kurokawa lift, then we use the relation (3.7), partial fractions and geometric series to show that $\lambda(m) > 0$ for all m.

Exercise 5.6 *Show that if $F \in S_k(\Gamma_2)$ is a Saito–Kurokawa lift, then $\lambda(m) > 0$ for all m.*

Theorem 5.7 (Kohnen [49]) *Let $F \in S_k(\Gamma_2)$ be in the orthogonal complement of the space of Saito–Kurokawa lifts. Suppose F is a Hecke eigenform with eigenvalues $\lambda(m)$ for $m \geq 1$. Then, there are infinitely many sign changes in the sequence $\{\lambda(m) : m \in \mathbb{N}\}$.*

Proof By Exercise 3.10, we have

$$\sum_{m \geq 1} \frac{\lambda(m)}{m^s} = \zeta(2s - 2k + 4)^{-1} L(s, F, \text{spin}), \qquad (5.3)$$

in the region where these converge. Suppose $\lambda(m)$ is positive for all $m \geq m_0$ for some $m_0 \in \mathbb{N}$. Landau's theorem [55, Page 536] states that either $\sum \lambda(m)m^{-s}$ converges for all $s \in \mathbb{C}$, or has a pole at the real point of its line of convergence.

We will first show that the former is true. Recall that we know from Theorem 3.11, for a non-Saito–Kurokawa lift F, the completed spin L-function

$$\Lambda(s, F) = (2\pi)^{-2s} \Gamma(s) \Gamma(s - k + 2) L(s, F, \text{spin})$$

is entire. The gamma function $\Gamma(s)$ has no zeros and has poles only at nonpositive integers. Hence, we can conclude that $L(s, F, \text{spin})$ is entire and has zeros at $s = k - 2, k - 3, \cdots$. Since $\zeta(s)$ has no zeros on the real line except at negative even integers (and these zeros are simple), the right-hand side of (5.3) has no poles on the real line. By Landau, this implies that $\sum \lambda(m)m^{-s}$ converges for all $s \in \mathbb{C}$.

Now, choose $s_0 \in \mathbb{C}$ such that $\text{Re}(2s_0 - 2k + 4) = 1/2$ and $\zeta(2s_0 - 2k + 4) = 0$. Hence, $L(s, F, \text{spin})$ has a zero at $s = s_0$. This implies $\Lambda(s_0, F) = 0$ as well. Once again, since F is not a Saito–Kurokawa lift, we know that F satisfies the generalized Ramanujan conjecture. Hence, the Satake p-parameters satisfy $|\alpha_{1,p}| = |\alpha_{2,p}| = 1$. Together with $\alpha_{0,p}^2 \alpha_{1,p} \alpha_{2,p} = p^{2k-3}$, we can conclude that the Euler product defining $L(s, F, \text{spin})$ converges absolutely for $\text{Re}(s) > k - 1/2$. This means that $L(s, F, \text{spin})$ is nonzero in the region $\text{Re}(s) > k - 1/2$. Hence, $\Lambda(s, F)$ is nonzero in the region $\text{Re}(s) > k - 1/2$. The functional equation (Theorem 3.11) implies $\Lambda(2k - 2 - s_0) = (-1)^k \Lambda(s_0, F) = 0$. But $\text{Re}(2k - 2 - s_0) = k - 1/4$, which leads to a contradiction.

Let us state some refinements of Kohnen's result.

(i) In 2007, Kohnen and Sengupta [52] showed that there exists $n \in \mathbb{N}$ satisfying

$$n \ll k^2 \log^{20} k$$

such that $\lambda(n) < 0$. In 2010, Jim Brown [15] obtained a similar result for Siegel cusp forms with level $N > 1$.

(ii) In 2008, Pitale and Schmidt [73] showed that there are infinitely many prime numbers p such that the sequence of Hecke eigenvalues $\{\lambda(p^r) : r \geq 0\}$ has infinitely many sign changes. This used representation theoretic techniques, and a weaker result than GRC.

(iii) Recently, in 2016, Royer et al. [84] showed that half of the nonzero coefficients in the Dirichlet series for the spin L-function of F are positive and half are negative.

For higher genus $n > 3$, we do not know about the meromorphic continuation and location of poles of the spin L-function. Also, GRC is still open for $n > 2$. Hence, we do not have any results on sign changes of Hecke eigenvalues in higher genus.

5.3   Sign Changes of Fourier Coefficients

Let $F \in S_k(\Gamma_n)$ with Fourier coefficients $\{A(T) : T > 0\}$. The Koecher–Maass series of F is given by

$$D(s) := \sum \frac{A(T)}{\epsilon(T) \det(T)^s},$$

where the sum is over all half-integral, symmetric, and positive definite T modulo the action of $GL_n(\mathbb{Z})$. Here $\epsilon(T) = \#\{U \in GL_n(\mathbb{Z}) : {}^tUTU = T\}$. The above series converges for $\mathrm{Re}(s) > (k + n + 1)/2$.

Theorem 5.8 (Maass [59, Page 217]) *Let $k > 2n$ and k be even. Let $F \in S_k(\Gamma_n)$ with Fourier coefficients $\{A(T) : T > 0\}$. Let*

$$\xi_F(s) := 2(2\pi)^{-ns} \prod_{i=1}^{n} \pi^{\frac{i-1}{2}} \Gamma(s - \frac{i-1}{2}) D(s).$$

Then $\xi_F(s)$ has a meromorphic continuation to all of $\mathbb{C}$ and satisfies the functional equation $\xi_F(k - s) = (-1)^{nk/2} \xi_F(s)$.

The key step in the proof of the above theorem is to realize $\xi_F(s)$ as a Mellin transform of F. Let us consider the $n = 2$ case. Let P_2 denote the space of all positive definite, symmetric 2×2 real matrices. Let R_2 be the Minkowski's reduced domain defined as the elements $Y = \begin{bmatrix} y_{11} & y_{12} \\ y_{21} & y_{22} \end{bmatrix} \in P_2$ such that the following conditions are satisfied.

(1) ${}^tgYg \geq y_{22}$ for all integral column matrices g,

(2) ${}^tgYg \geq y_{11}$ for all integral column matrices $g = \begin{bmatrix} g_1 \\ g_2 \end{bmatrix}$ with $(g_1, g_2) = 1$,

(3) $y_{12} \geq 0$.

See Page 12 of [46] for further details. We have the following relation between R_2 and P_2:

$$P_2 = \sqcup {}^tgR_2g$$

where the union is over all $g \in GL_2(\mathbb{Z})$.

Exercise 5.9 *Let R_2 and P_2 be as above.*

(1) *Substitute the Fourier expansion of F, and use the relation between P_2 and R_2, to show that*

$$\int_{R_2} F(iY) \det Y^{s-3/2} dY = \sum_{\{T\}>0} \frac{A(T)}{\epsilon(T)} \int_{P_2} \det Y^{s-3/2} e^{-2\pi \operatorname{Tr}(TY)} dY.$$

(2) *We can assume that T is a diagonal matrix given by $\begin{bmatrix} t_1 & \\ & t_2 \end{bmatrix}$. Write*

$$Y = \begin{bmatrix} 1 & \\ y_3 & 1 \end{bmatrix} \begin{bmatrix} y_1 & \\ & y_2 \end{bmatrix} \begin{bmatrix} 1 & y_3 \\ & 1 \end{bmatrix}.$$

Show that

$$\int_{P_2} \det Y^{s-3/2} e^{-2\pi \operatorname{Tr}(TY)} dY$$

$$= \int_{y_1>0, y_2>0, y_3} y_1^{s-1/2} e^{-2\pi y_1 t_1} y_2^{s-3/2} e^{-2\pi y_2 t_2} e^{-2\pi y_3^2 y_1 t_2} dy_1 dy_2 dy_3.$$

(3) *Use*

$$\int_0^\infty y^{s-1} e^{-y} dy = \Gamma(s), \qquad \int_{-\infty}^\infty e^{-y^2} dy = \frac{2}{\sqrt{\pi}}$$

to conclude that

$$\xi_F(s) = \int_{R_2} F(iY) \det Y^{s-3/2} dY.$$

The Koecher–Maass series was used by Jesgarz [44] to show the following: Suppose $F \in S_k(\Gamma_n)$ have real Fourier coefficients. Then, there are infinitely many $T > 0$ (modulo the action of $\operatorname{GL}_n(\mathbb{Z})$) with $A(T) > 0$, and similarly such that $A(T) < 0$.

Theorem 5.10 (Choie, Gun, Kohnen [18]) *Let k be even and let $F \in S_k(\Gamma_n)$, $F \neq 0$. Suppose that the Fourier coefficients $\{A(T) : T > 0\}$ are real. Then, there exist $T_1 > 0, T_2 > 0$ with*

$$\operatorname{Tr}(T_1), \operatorname{Tr}(T_2) \ll (k \cdot c_n)^5 \log^2 6(k \cdot c_n),$$

such that $A(T_1) > 0$ and $A(T_2) < 0$. Here

$$c_n := n2^{n-1}(4/3)^{n(n-1)/2}.$$

The main idea is to look at the Fourier–Jacobi expansion of F where the coefficients are Jacobi forms. Using Taylor expansions of these coefficients the authors reduce the question to the case of elliptic modular forms, and then apply the results of Choie and Kohnen in [19].

Chapter 6
Cuspidal Automorphic Representations Corresponding to Siegel Modular Forms

In this chapter, we start with a cuspidal Hecke eigenform $F \in S_k(\Gamma_n)$ and construct an irreducible cuspidal automorphic representation of $\mathrm{GSp}_{2n}(\mathbb{A})$ corresponding to it. There are several steps for achieving this—construct a function Φ_F on $\mathrm{GSp}_{2n}(\mathbb{A})$ corresponding to F, understand the properties it inherits from F, and study the local components of the representation generated by Φ_F. The main reference for this chapter is the article [6] by Asgari and Schmidt. We suggest the reader to go over Appendix B and C to refresh the details about adeles and local representation theory of GL_2.

6.1 Classical to Adelic

For convenience, let us now denote by G the group GSp_{2n}. Let $F \in S_k(\Gamma_n)$. We will define a function Φ_F on $G(\mathbb{A})$, where $\mathbb{A}$ is the ring of adeles of $\mathbb{Q}$. We will use the decomposition of $G(\mathbb{A})$ given in the following exercise.

Exercise 6.1 *Use strong approximation of* Sp_{2n} *(see [47]) to get*

$$G(\mathbb{A}) = G(\mathbb{Q})G(\mathbb{R})^+ \prod_{p<\infty} G(\mathbb{Z}_p).$$

Let us denote $G_\infty = G(\mathbb{R})$. Write an element $g \in G(\mathbb{A})$ as

$$g = g_\mathbb{Q} g_\infty k_0, \quad \text{where } g_\mathbb{Q} \in G(\mathbb{Q}),\, g_\infty \in G_\infty^+,\, k_0 \in K_0,$$

where $K_0 = \prod_{p<\infty} K_p$, with $K_p = G(\mathbb{Z}_p)$. Then define

$$\Phi_F(g) := (F\|_k g_\infty)(I), \tag{6.1}$$

© Springer Nature Switzerland AG 2019
A. Pitale, *Siegel Modular Forms*, Lecture Notes in Mathematics 2240,
https://doi.org/10.1007/978-3-030-15675-6_6

where $I = i1_n$. The action $\|_k$ is defined by

$$(F\|_k g)(Z) := \mu(g)^{nk/2} \det J(g, Z)^{-k} F(g\langle Z \rangle), \quad g \in G(\mathbb{R})^+, \; Z \in \mathbb{H}_n. \quad (6.2)$$

Note that, this slash action is different from the one we defined in (3.2) for the classical setting. The relation between the two slash actions is given by

$$F|_k g = \mu(g)^{nk/2 - n(n+1)/2} F\|_k g. \quad (6.3)$$

The reason for the two slash actions is that, the classical $|_k$ works very well to give a nice formula for the Dirichlet series, while $\|_k$ defined above ensures that Φ_F is trivial under the center of G (see below).

Exercise 6.2 *In this exercise, we will show that Φ_F is well defined.*

(i) *Consider an element $g = g_\infty \otimes_{p<\infty} 1 \in G(\mathbb{A})$, with $g_\infty \in G_\infty^+$. Suppose $g = g' := g'_\mathbb{Q} g'_\infty k'_0$ is another expression for g. Show that $g_\infty = g'_\mathbb{Q} g'_\infty$ and $g'_\mathbb{Q} \in \Gamma_n$.*

(ii) *Use the automorphy of F to show that Φ_F is well defined.*

The map $F \mapsto \Phi_F$ gives an injection from $S_k(\Gamma_n)$ to a space of functions $\Phi :$ $G(\mathbb{A}) \to \mathbb{C}$ satisfying the following properties.

(i) $\Phi(\gamma g) = \Phi(g)$ for $\gamma \in G(\mathbb{Q})$,
(ii) $\Phi(g k_0) = \Phi(g)$ for $k_0 \in K_0$,
(iii) $\Phi(g k_\infty) = \det J(k_\infty, I)^{-k} \Phi(g)$ for $k_\infty \in K_\infty$,
(iv) $\Phi(gz) = \Phi(g)$ for $z \in Z(\mathbb{A})$.

Here $Z \cong \mathrm{GL}_1$ is the center of G, and $K_\infty \simeq U_n$ is the standard maximal compact subgroup of $\mathrm{Sp}_{2n}(\mathbb{R})$. We can see that K_∞ is the stabilizer of I.

Lemma 6.3 *Let $F \in S_k(\Gamma_n)$. Then Φ_F is cuspidal, i.e., for every unipotent radical N of every proper parabolic subgroup of G, we have*

$$\int_{N(\mathbb{Q})\backslash N(\mathbb{A})} \Phi_F(ng)dn = 0, \text{ for all } g \in G(\mathbb{A}).$$

Note that we are using n for the genus, as well as elements of the unipotent subgroup N. Hopefully, the context makes the choice clear, and there is no confusion.

Remark 6.4 For $n = 2$, there are 3 standard proper parabolic subgroups of G.

(1) The Borel parabolic subgroup $B = \{ \begin{bmatrix} * & & * & * \\ * & * & * & * \\ & & * & * \\ & & & * \end{bmatrix} \in G\}$ with unipotent radical

$\{ \begin{bmatrix} 1 & & * & * \\ * & 1 & * & * \\ & & 1 & * \\ & & & 1 \end{bmatrix} \in B\}$.

(2) The Siegel parabolic subgroup $P = \{\begin{bmatrix} A & B \\ C & D \end{bmatrix} \in G : C = 0\}$ with unipotent radical $\{\begin{bmatrix} 1 & X \\ & 1 \end{bmatrix} : X \in M_2, \, {}^t X = X\}$.

(3) The Klingen parabolic subgroup $Q = \{\begin{bmatrix} * & * & * & \\ * & * & * & * \\ * & & * & * \\ & & & * \end{bmatrix} \in G\}$ with unipotent radical

$$\{\begin{bmatrix} 1 & & * & \\ * & 1 & * & * \\ & & 1 & * \\ & & & 1 \end{bmatrix} \in Q\}.$$

Sketch of proof of Lemma 6.3: It is enough to verify the cusp condition for the standard maximal parabolics. Let $P = MN$ be one of those. By the Iwasawa decomposition $G = PK$, with $K = K_0 K_\infty$, we can reduce to $g \in P(\mathbb{A})$. Since M normalizes N, we can reduce to $g \in M(\mathbb{A})$. Using strong approximation for $M(\mathbb{A})$, we can further assume that $g \in M_\infty^+ := M(\mathbb{R}) \cap G_\infty^+$. It will be enough if we can show that the integral over $V(\mathbb{Q}) \backslash V(\mathbb{A})$ is zero, where V is the intersection of all the unipotent radicals of all the maximal parabolic subgroups. (This is because, we can first integrate over the smaller space, and if the inner integral is zero, then the whole integral is zero.) Now, observe that V consists of matrices of the form $\begin{bmatrix} 1_n & X \\ & 1_n \end{bmatrix}$ with X in V', the set of symmetric $n \times n$ matrices with nonzero entries only in the last row and column. (Check this for the $n = 2$ case). With $Z = g\langle I \rangle$, we now get

$$\int_{V(\mathbb{Q}) \backslash V(\mathbb{A})} \Phi_F(ng) dn = \int_{V(\mathbb{Z}) \backslash V(\mathbb{R})} \Phi_F(ng) dn = \mu(g)^{nk/2} \int_{V(\mathbb{Z}) \backslash V(\mathbb{R})} F(ng\langle I \rangle) J(ng, I)^{-k} dn$$

$$= \mu(g)^{nk/2} \int_{V'(\mathbb{Z}) \backslash V'(\mathbb{R})} F(Z + X) J(g, I)^{-k} dX.$$

We have the Fourier expansion of F given by

$$F(Z) = \sum_{T > 0} A(T) e^{2\pi i \operatorname{Tr}(TZ)}.$$

We are only summing over $T > 0$ because F is a cusp form. Using this we get

$$\int_{V(\mathbb{Q}) \backslash V(\mathbb{A})} \Phi_F(ng) dn = \mu(g)^{nk/2} \sum_{T > 0} A(T) e^{2\pi i \operatorname{Tr}(TZ)} J(g, I)^{-k} \int_{V'(\mathbb{Z}) \backslash V'(\mathbb{R})} e^{2\pi i \operatorname{Tr}(TX)} dX.$$

Since T is nondegenerate, the map

$$X \mapsto e^{2\pi i \operatorname{Tr}(TX)}$$

is a nontrivial character on $V'(\mathbb{Z})\backslash V'(\mathbb{R})$. Hence, the integral is zero.   □

The next exercise shows how one can take Fourier coefficients of the adelic function Φ_F, and its relation to the Fourier coefficients of F.

Exercise 6.5 *Let $n = 2$. Let U be the unipotent radical of the Siegel parabolic subgroup of G. Let $\psi : \mathbb{Q}\backslash\mathbb{A} \to \mathbb{C}^\times$ be the character such that $\psi(x) = e^{2\pi i x}$ if $x \in \mathbb{R}$, and $\psi(x) = 1$ for $x \in \mathbb{Z}_p$. Given $S \in \mathrm{Sym}_2(\mathbb{Q})$, one obtains a character θ_S of $U(\mathbb{Q})\backslash U(\mathbb{A})$ by*

$$\theta_S(\begin{bmatrix} 1 & X \\ & 1 \end{bmatrix}) = \psi(\mathrm{Tr}(SX)).$$

Note that every character of $U(\mathbb{Q})\backslash U(\mathbb{A})$ is obtained in this way. For $S \in \mathrm{Sym}_2(\mathbb{Q})$ we define the following adelic Fourier coefficient of Φ_F,

$$\Phi_F^S(g) := \int\limits_{U(\mathbb{Q})\backslash U(\mathbb{A})} \Phi_F(ng)\theta_S^{-1}(n)\,dn \quad for\ g \in G(\mathbb{A}).$$

Show that, if $S \in \mathrm{Sym}_2(\mathbb{Q})$ is positive definite and half-integral, then

$$\Phi_F^S(1) = A(S)e^{-2\pi\,\mathrm{Tr}(S)}.$$

We have $\Gamma_n\backslash\mathbb{H}_n \simeq Z(\mathbb{A})G(\mathbb{Q})\backslash G(\mathbb{A})/K$. The Haar measure on $G(\mathbb{A})$ induces a measure on the right-hand side, and the corresponding measure on the left is a scalar multiple of the invariant volume element d^*Z on $\mathbb{H}_n$. In other words, if a Γ_n-invariant function f on $\mathbb{H}_n$ and a $G(\mathbb{Q})$-invariant function Φ on $Z(\mathbb{A})\backslash G(\mathbb{A})/K$ are related by $f(g_\infty\langle I\rangle) = \Phi(g_\infty)$, for all $g_\infty \in G_\infty^+$, then

$$\int\limits_{\Gamma_n\backslash\mathbb{H}_n} f(Z)d^*Z = \int\limits_{Z(\mathbb{A})G(\mathbb{Q})\backslash G(\mathbb{A})/K} \Phi(g)dg = \int\limits_{Z(\mathbb{A})G(\mathbb{Q})\backslash G(\mathbb{A})} \Phi(g)dg.$$

Exercise 6.6 *Show that, if $F_1, F_2 \in S_k(\Gamma_n)$, then*

$$\langle F_1, F_2\rangle = \langle \Phi_{F_1}, \Phi_{F_2}\rangle.$$

Here, on the left-hand side we have the Petersson inner product (1.10) of classical Siegel modular forms, and on the right-hand side, we have the ordinary L^2-scalar product given by $\int_{Z(\mathbb{A})G(\mathbb{Q})\backslash G(\mathbb{A})} \Phi_{F_1}(g)\overline{\Phi_{F_2}(g)}dg$.

6.2 Hecke Equivariance

Let $G_p = G(\mathbb{Q}_p)$ and $K_p = G(\mathbb{Z}_p)$. Let $\mathcal{H}(G_p, K_p)$ be the Hecke algebra of G_p consisting of compactly supported functions $f : G_p \to \mathbb{C}$, which are left and right K_p-invariant. The product in $\mathcal{H}(G_p, K_p)$ is given by convolution

$$(f * g)(x) = \int_{G_p} f(xy)g(y^{-1})dy.$$

Exercise 6.7 $\mathcal{H}(G_p, K_p)$ is spanned by the characteristic functions of the sets $K_p g K_p$ for $g \in G_p$.

Recall that $\mathcal{H}_{n,p}$ is the p-component of the classical Hecke algebra defined in Sect. 3.1. We know that $\mathcal{H}_{n,p}$ is generated by double cosets $\Gamma_n M \Gamma_n$ for $M \in G(\mathbb{Z}[p^{-1}])^+$. Here, $\mathbb{Z}[p^{-1}]$ is the ring of rational numbers with only powers of p in the denominator. There is a bijection $\Gamma_n \backslash G(\mathbb{Z}[p^{-1}])^+ / \Gamma_n \simeq G(\mathbb{Z}) \backslash G(\mathbb{Z}[p^{-1}])^+ / G(\mathbb{Z}) \simeq K_p \backslash G_p / K_p$ induced by the inclusions of the groups. This gives us an isomorphism

$$\mathcal{H}_{n,p} \simeq \mathcal{H}(G_p, K_p).$$

We will henceforth identify these two Hecke algebras. The definition (3.1) gives us the action of $\mathcal{H}_{n,p}$ on $S_k(\Gamma_n)$. We can also define an action of $\mathcal{H}(G_p, K_p)$ on adelic functions as follows.

$$(T\Phi)(g) = \int_{G_p} T(h)\Phi(gh)dh, \qquad T \in \mathcal{H}(G_p, K_p), g \in G(\mathbb{A}).$$

If T is the characteristic function of $K_p M K_p = \sqcup_i M_i K_p$ and Φ is right K_p invariant then

$$(T\Phi)(g) = \sum_i \Phi(gM_i).$$

Here, we are assuming that the Haar measure on G_p is normalized so that the volume of K_p is 1. If $T = \Gamma_n M \Gamma_n = \sqcup_i \Gamma_n M_i'$, then we can define

$$F||_k T = \sum_i F||_k M_i'.$$

From (6.3), it is clear that $F||_k T = \mu(M)^{n(n+1-k)/2} T F$, where $T F$ is the action of the Hecke operator on F using the $|_k$ action.

Lemma 6.8 *Identify the Hecke algebras $\mathcal{H}_{n,p}$ and $\mathcal{H}(G_p, K_p)$. Let $F \in S_k(\Gamma_n)$ and Φ_F be the corresponding function on $G(\mathbb{A})$. For every $T \in \mathcal{H}(G_p, K_p)$, we have*

$$T\Phi_F = \Phi_{F\|_k T}.$$

This is Lemma 9 of [6]. We can now summarize the results of this section in the following theorem.

Theorem 6.9 *The mapping $F \mapsto \Phi_F$ given in (6.1) maps the space $S_k(\Gamma_n)$ of classical Siegel cusp forms of weight k and degree n isometrically and in a Hecke equivariant way into a subspace of $L_0^2(Z(\mathbb{A})G(\mathbb{Q})\backslash G(\mathbb{A}))$ consisting of continuous functions Φ on $G(\mathbb{A})$ with the following properties:*

- *(i) $\Phi(\gamma g) = \Phi(g)$ for $\gamma \in G(\mathbb{Q})$*
- *(ii) $\Phi(g k_0) = \Phi(g)$ for $k_0 \in K_0$*
- *(iii) $\Phi(g k_\infty) = J(k_\infty, I)^{-k}\Phi(g)$ for $k_\infty \in K_\infty \simeq U_n$*
- *(iv) $\Phi(gz) = \Phi(g)$ for $z \in Z(\mathbb{A})$*
- *(v) Φ is a smooth function on $G(\mathbb{R})^+$ (fixed finite components), and is annihilated by $\mathfrak{p}_{\mathbb{C}}^{-}$,*
- *(vi) Φ is cuspidal.*

We have added part (v) in the above theorem, which provides information on the archimedean component. See Lemma 7 of [6] for details.

6.3 Satake Isomorphism

Let us briefly discuss the Satake isomorphism for Hecke algebras. The standard maximal torus of G is given by

$$T := \{\mathrm{diag}(u_1, u_2, \cdots, u_n, u_1^{-1}u_0, u_2^{-1}u_0, \cdots, u_n^{-1}u_0) : u_i \in \mathrm{GL}_1 \quad \text{for } i = 0, 1, \cdots, n\}.$$

Let $T^\circ := T(\mathbb{Z}_p) \subset T(\mathbb{Q}_p)$. Consider the Hecke algebra $\mathcal{H}(T, T^\circ)$ of compactly supported functions on T that are right and left T°-invariant. Consider the $n + 1$ elements of $\mathcal{H}(T, T^\circ)$ given by

$$X_0 := \mathrm{char}(\mathrm{diag}(\mathbb{Z}_p^\times, \cdots, \mathbb{Z}_p^\times, p\mathbb{Z}_p^\times, \cdots, p\mathbb{Z}_p^\times)),$$
$$X_1 := \mathrm{char}(\mathrm{diag}(p\mathbb{Z}_p^\times, \mathbb{Z}_p^\times, \cdots, \mathbb{Z}_p^\times, p^{-1}\mathbb{Z}_p^\times, \mathbb{Z}_p^\times, \cdots, \mathbb{Z}_p^\times)),$$
$$\cdots X_n := \mathrm{char}(\mathrm{diag}(\mathbb{Z}_p^\times, \cdots, \mathbb{Z}_p^\times, p\mathbb{Z}_p^\times, \mathbb{Z}_p^\times, \cdots, \mathbb{Z}_p^\times, p^{-1}\mathbb{Z}_p^\times)).$$

Exercise 6.10 *Show that*

$$\mathcal{H}(T, T^\circ) = \mathbb{C}[X_0^{\pm 1}, X_1^{\pm 1}, \cdots, X_n^{\pm 1}].$$

For an element $f \in \mathcal{H}(G_p, K_p)$, define the Satake transform by

$$(Sf)(t) := |\delta_B(t)|^{\frac{1}{2}} \int_N f(tn)dt = |\delta_B(t)|^{-\frac{1}{2}} \int_N f(nt)dt,$$

where

$$\delta_B(t) = u_0^{-n(n+1)/2} u_1^2 u_2^4 \cdots u_n^{2n}, \text{ for } t = \operatorname{diag}(u_1, u_2, \cdots, u_n, u_1^{-1} u_0, u_2^{-1} u_0, \cdots, u_n^{-1} u_0).$$

Here, N is the unipotent radical of the Borel subgroup B. It consists of matrices of the form $\begin{bmatrix} A & \\ & {}^t A^{-1} \end{bmatrix} \begin{bmatrix} 1 & X \\ & 1 \end{bmatrix}$, where $A \in GL_n(\mathbb{Q}_p)$ is lower triangular with 1's on the diagonal, and X is a symmetric matrix in $M_n(\mathbb{Q}_p)$. The Weyl group W of G acts on the torus T by permuting $u_1, \cdots, u_n$ and replacing u_i's by $u_0 u_i^{-1}$. It can be shown that $Sf \in \mathcal{H}(T, T^\circ)^W$, the space of W-invariant elements. In fact, the Satake map gives an isomorphism (see Section 3.1 of [6])

$$\mathcal{H}(G_p, K_p) \simeq \mathcal{H}(T, T^\circ)^W.$$

Let us see the image of this map on the characteristic function of a double coset $K_p M K_p$. Suppose $K_p M K_p = \sqcup_i M_i K_p$, where

$$M_i = \begin{bmatrix} A_i & B_i \\ & p^{d_{i0}} {}^t A_i^{-1} \end{bmatrix}, \text{ with } A_i = \begin{bmatrix} p^{d_{i1}} & & 0 \\ & \ddots & \\ * & & p^{d_{in}} \end{bmatrix}.$$

Here d_{i0} does not depend on i, it equals the valuation of $\mu(M)$.

Let $t = \operatorname{diag}(p^{k_1}, \cdots, p^{k_n}, p^{k_0 - k_1}, \cdots, p^{k_0 - k_n})$. We have

$$\int_N \mathbf{1}_{M_i K_p}(tn)dn = \begin{cases} 1 & \text{if } k_j = d_{ij} \text{ for all } j = 0, 1, \cdots, n, \\ 0 & \text{otherwise.} \end{cases}$$

Here, $\mathbf{1}_X$ is the characteristic function of the set X. We can conclude that

$$\left(t \mapsto \int_N \mathbf{1}_{M_i K_p}(tn)dn \right) = X_0^{d_{i0}} X_1^{d_{i1}} \cdots X_n^{d_{in}}.$$

This leads to the following formula for the Satake transform.

Lemma 6.11 *Let* $f = \operatorname{char}(\Gamma_n M \Gamma_n)$, *with* M *as above. Let* δ *be the valuation of* $\mu(M)$. *Then*

$$Sf = p^{\delta n(n+1)/4} X_0^\delta \sum_i \prod_{j=1}^n (p^{-j} X_j)^{d_{ij}}.$$

This is Lemma 1 of [6]. We will compare this with the classical formula (3.3) in Exercise 6.13.

6.4 Spherical Representations

An irreducible, admissible representation of G_p is called *spherical* if it contains a nonzero vector fixed by K_p. All the spherical representations of G_p are obtained as follows. Let $\chi_0, \chi_1, \cdots, \chi_n$ be unramified characters of $\mathbb{Q}_p^\times$ (i.e., homomorphisms from $\mathbb{Q}_p^\times \to \mathbb{C}^\times$ that are trivial on $\mathbb{Z}_p^\times$). They define an unramified character, call it χ, of the Borel subgroup $B = TN$ which is trivial on N and, on T, is given by

$$t = \operatorname{diag}(u_1, u_2, \cdots, u_n, u_1^{-1}u_0, u_2^{-1}u_0, \cdots, u_n^{-1}u_0) \mapsto \chi_0(u_0)\chi_1(u_1) \cdots \chi_n(u_n).$$

The representation $\operatorname{Ind}_B^{G_p}(\chi)$ of G_p obtained by normalized induction from χ consists of locally constant functions on G_p satisfying

$$f(tng) = |\delta_B(t)|^{1/2}\chi(t)f(g), \quad \text{for all } t \in T, n \in N \text{ and } g \in G_p.$$

Exercise 6.12 *Show that the central character of the above induced representation is* $\chi_0^2\chi_1 \cdots \chi_n$.

The representation $\operatorname{Ind}_B^{G_p}(\chi)$ has a unique sub-representation or a sub-quotient which is a spherical representation. It is called the spherical constituent and let us denote it by

$$\pi = \pi(\chi_0, \chi_1, \cdots, \chi_n).$$

The isomorphism class of this representation depends only on the unramified characters modulo the action of the Weyl group. It is further known that every spherical representation of G_p is obtained in this way. Thus, there is a bijection between unramified characters of T modulo the action of the Weyl group, and isomorphism classes of spherical representations of G_p. Each unramified character of $\mathbb{Q}_p^\times$ is determined by its value on p. This value may be any nonzero complex number. Hence, any vector $(b_0, b_1, \cdots, b_n) \in (\mathbb{C}^\times)^{n+1}/W$ gives a character χ, up to action by the Weyl group. We have the following diagram.

$$\{ \text{spherical representations} \} \longrightarrow \operatorname{Hom}_{\mathrm{alg}}(\mathcal{H}(G_p, K_p), \mathbb{C})$$

$$\{ \text{unramified characters} \}/W \longleftarrow (\mathbb{C}^\times)^{n+1}/W$$

All the maps above are bijections. The left arrow is induction and then taking the spherical constituent, the top arrow is the action of $\mathcal{H}(G_p, K_p)$ on the spherical vector, the map on the right comes from the identification $\mathcal{H}(G_p, K_p) \simeq \mathbb{C}[X_0^{\pm 1}, X_1^{\pm 1}, \cdots, X_n^{\pm 1}]^W$, and the bottom arrow assigns to the vector $(b_0, b_1, \cdots, b_n)$ the characters $\chi_i(p) = b_i$.

Finally, we have the following local standard and spin L-functions of the representation π:

$$L(s, \pi, \text{std})^{-1} := (1 - p^{-s}) \prod_{i=1}^{n} (1 - b_i p^{-s})(1 - b_i^{-1} p^{-s})$$

$$L(s, \pi, \text{spin})^{-1} := \prod_{k=0}^{n} \prod_{1 \leq i_1 < \cdots < i_k \leq n} (1 - b_0 b_{i_1} \cdots b_{i_k} p^{-s})$$

See Section 3.2 of [6] for further details on spherical representations of G_p.

6.5 The Representation Associated to a Siegel Cusp Form

Let $F \in S_k(\Gamma_n)$ be a Hecke eigenform and let Φ_F be the corresponding cusp form on $G(\mathbb{A})$. For any $g \in G(\mathbb{A})$, the translate of Φ_F by g is the function $G(\mathbb{A}) \ni h \mapsto \Phi_F(hg)$. Let V_F be the subspace of $L_0^2(Z(\mathbb{A})G(\mathbb{Q})\backslash G(\mathbb{A}))$ spanned by the translates of Φ_F. The group $G(\mathbb{A})$ acts on V_F by right translation giving us a represenattion π_F of $G(\mathbb{A})$. A result of Narita, Pitale, Schmidt [64, Theorem 3.1] guarantees that π_F is irreducible. π_F is a cuspidal automorphic representation of $G(\mathbb{A})$, which is trivial on $Z(\mathbb{A})$. We may thus consider π_F as an automorphic representation of $\text{PGSp}_{2n}(\mathbb{A})$.

Let $\pi_F \cong \otimes_p \pi_p$, be the restricted tensor product. Here, π_p is an irreducible admissible representation of G_p. For a finite prime p, Theorem 6.9 tells us that Φ_F is right K_p-invariant. Hence, for every finite prime p, the representation π_p is a spherical representation of G_p. Hence, π_p is of the form $\pi(\chi_0, \chi_1, \cdots, \chi_n)$ for some unramified characters χ_i of $\mathbb{Q}_p^{\times}$.

Exercise 6.13 *Set $b_i = \chi_i(p)$ as before. Recall $\alpha_{0,p}, \alpha_{1,p}, \cdots, \alpha_{n,p}$, the classical Satake p-parameters of F defined in (3.3).*

(i) *Show that*

$$b_0 = p^{n(n+1)/4 - nk/2} \alpha_{0,p} \text{ and } b_i = \alpha_{i,p} \text{ for } i = 1, \cdots, n.$$

(ii) *Show that π_p has trivial central character.*

(iii) *Let $L(s, \pi_F, \text{spin}) = \prod_p L(s, \pi_p, \text{spin})$. Find the relation between $L(s, F, \text{spin})$ and $L(s, \pi_F, \text{spin})$. Further, for $n = 2$, find out what kind of functional equation does the completion of $L(s, \pi_F, \text{spin})$ satisfy.*

At the archimedean place, π_∞ is a representation of $G(\mathbb{R})$ containing a vector v_∞ such that $\mathfrak{p}_{\mathbb{C}}^- v_\infty = 0$ and, $\pi_\infty(k_\infty) v_\infty = J(k_\infty, I)^{-k} v_\infty$ for every $k_\infty \in K_\infty$. A representation of $G(\mathbb{R})$ has such a vector if and only if it is a *lowest weight representation* π_k of $G(\mathbb{R})$. It is in the holomorphic discrete series if $k > n$, and is in the limit of a holomorphic discrete series if $k = n$.

Theorem 6.14 *Let $F \in S_k(\Gamma_n)$ be a Hecke eigenform and let π_F be the cuspidal automorphic representation of* $\mathrm{PGSp}_{2n}(\mathbb{A})$ *associated by F. Let $\pi_F \cong \otimes_p \pi_p$.*

(i) *For a finite prime p, the local representation π_p is a spherical representation of* $\mathrm{PGSp}_{2n}(\mathbb{Q}_p)$ *with unramified characters $\chi_0, \chi_1, \cdots, \chi_n$ given by $\chi_i(p) = \alpha_{i,p}$ for all $i = 1, \cdots, n$ and $\chi_0(p) = p^{n(n+1)/4 - nk/2} \alpha_{0,p}$, where the $\alpha_{0,p}, \cdots, \alpha_{n,p}$ are the classical Satake p-parameters of F.*

(ii) *At the archimedean place, π_∞ is the lowest weight representation π_k of $G(\mathbb{R})$.*

Chapter 7
Local Representation Theory
of $\mathrm{GSp}_4(\mathbb{Q}_p)$

In this chapter, we will discuss the representation theory of $\mathrm{GSp}_4(\mathbb{Q}_p)$. The genus 2 case provides a very good introduction to the study of local representations of the symplectic groups. The genus 2 case also has the advantage of detailed tables of data compiled by Roberts and Schmidt. We will present some of these tables and data corresponding to the classification of the local representations, the Iwahori-fixed vectors in the Iwahori-spherical representations and the paramodular theory.

7.1 Local Non-archimedean Representations for GSp_4

In the previous chapter, we have seen that the local non-archimedean component π_p of the automorphic representation π_F associated to a Hecke eigenform $F \in S_k(\Gamma_n)$ is a spherical representation. This is obtained from a representation of $\mathrm{GSp}_{2n}(\mathbb{Q}_p)$ induced from an unramified character of the Borel subgroup. We will try to understand how a general irreducible admissible representation of $\mathrm{GSp}_{2n}(\mathbb{Q}_p)$ looks like. We will restrict ourselves to genus 2. As before, let us denote $G = \mathrm{GSp}_4$. Let $\pi \cong \otimes_p \pi_p$ be an irreducible cuspidal automorphic representation of $G(\mathbb{A})$. For a finite prime p, each π_p is an irreducible admissible representation of $G_p = G(\mathbb{Q}_p)$. There are two possibilities for π_p—it is either supercuspidal or not. In the latter case, we know that it is obtained as a constituent of an induced representation. In the previous section, we have already seen representations induced from the Borel subgroup. As we saw in Remark 6.4, GSp_4 has two more proper parabolic subgroups. Let us set up notation for the representations induced from these parabolic subgroups now.

(i) *Borel subgroup B*: Let χ_1, χ_2, σ be characters of $\mathbb{Q}_p^\times$. As before, we get a character of the Borel subgroup $B(\mathbb{Q}_p)$ given by

© Springer Nature Switzerland AG 2019

A. Pitale, *Siegel Modular Forms*, Lecture Notes in Mathematics 2240,
https://doi.org/10.1007/978-3-030-15675-6_7

$$\begin{bmatrix} a & * & & * \\ * & b & * & & * \\ & & ca^{-1} & * \\ & & & cb^{-1} \end{bmatrix} \mapsto \chi_1(b)\chi_2(a)\sigma(c).$$

The representation of G_p obtained by normalized parabolic induction of this character of $B(\mathbb{Q}_p)$ is denoted by $\chi_1 \times \chi_2 \rtimes \sigma$. The standard model of this representation consists of all locally constant functions $f : G_p \to \mathbb{C}$ that satisfy

$$f(\begin{bmatrix} a & * & & * \\ * & b & * & & * \\ & & ca^{-1} & * \\ & & & cb^{-1} \end{bmatrix} g) = |ab^2||c|^{-3/2}\chi_1(b)\chi_2(a)\sigma(c)f(g).$$

The group acts on this space by right translation and the central character of $\chi_1 \times \chi_2 \rtimes \sigma$ is $\sigma^2\chi_1\chi_2$.

(ii) *Siegel parabolic subgroup P*: Let (π, V) be an irreducible admissible representation of $GL_2(\mathbb{Q}_p)$ and let σ be a character of $\mathbb{Q}_p^\times$. Then, we denote by $\pi \rtimes \sigma$ the representation of G_p obtained by normalized parabolic induction from the representation of the Siegel parabolic subgroup $P(\mathbb{Q}_p)$ on V given by

$$\begin{bmatrix} A & * \\ & c\,{}^tA^{-1} \end{bmatrix} \mapsto \sigma(c)\pi(A).$$

The standard model of this representation consists of all locally constant functions $f : G_p \to V$ that satisfy

$$f(\begin{bmatrix} A & * \\ & c\,{}^tA^{-1} \end{bmatrix} g) = |\det(A)c^{-1}|^{3/2}\sigma(c)\pi(A)f(g).$$

If ω_π is the central character of π, then the central character of $\pi \rtimes \sigma$ is $\omega_\pi\sigma^2$.

(iii) *Klingen parabolic subgroup Q*: Let χ be a character of $\mathbb{Q}_p^\times$ and let (π, V) be an irreducible admissible representation of $GL_2(\mathbb{Q}_p)$. Then, we denote by $\chi \rtimes \pi$ the representation of G_p obtained by normalized parabolic induction from the representation of the Klingen parabolic $Q(\mathbb{Q}_p)$ on V given by

$$\begin{bmatrix} a & b & & * \\ * & t & * & & * \\ c & d & & * \\ & & & \Delta t^{-1} \end{bmatrix} \mapsto \chi(t)\pi(\begin{bmatrix} a & b \\ c & d \end{bmatrix}), \qquad \Delta = ad - bc.$$

The standard model of this representation consists of all locally constant functions $f : G_p \to V$ that satisfy

$$f\left(\begin{bmatrix} a & b & * & \\ * & t & * & * \\ c & d & * \\ & & & \Delta t^{-1} \end{bmatrix} g\right) = |t^2(ad - bc)^{-1}|\chi(t)\pi\left(\begin{bmatrix} a & b \\ c & d \end{bmatrix}\right)f(g).$$

If ω_π is the central character of π, then the central character of $\chi \rtimes \pi$ is $\chi\omega_\pi$.

Exercise 7.1 *Let $\pi = \chi_1 \times \chi_2 \rtimes \sigma$ be the Borel induced representation of G_p. There are 8 elements in the Weyl group W of G_p. Given an element $w \in W$, we can define an action on the torus as in Sect. 6.3, and get a new character from χ_1, χ_2, σ, which gives another Borel-induced representation π^w of G_p. It is known that the constituents of both π and π^w are the same. In particular, if they are irreducible, then $\pi \simeq \pi^w$. List all the possible π^w for $w \in W$.*

Using the results of Sally and Tadic [90], we describe a useful listing of the non-supercuspidal, irreducible, and admissible representations of G_p. The basis for this list is the fact that every non-supercuspidal, irreducible, and admissible representation of G_p is a constituent (irreducible subquotient) of a parabolically induced representation with proper supercuspidal inducing data. Given this, one might try to classify the non-supercuspidal, irreducible, and admissible representations of G_p by doing the following: First, write down all the supercuspidal inducing data for the Borel, Klingen, and Siegel parabolic subgroups of G_p. Second, determine the Langlands classification data of all the constituents of all the resulting parabolically induced representations. Third, find all the possible ways in which a fixed non-supercuspidal, irreducible, and admissible representation of G_p arises as a constituent in the second step; the results of this step may show that some of the supercuspidal inducing data from the first step is redundant from the point of view of listing all representations.

In fact, the paper of Sally and Tadic [90] has carried out the difficult aspects of this procedure. It turns out that eleven groups of supercuspidal inducing data are required. Groups I–VI contain representations supported in B, i.e., these representations are constituents of induced representations of the form $\chi_1 \times \chi_2 \rtimes \sigma$. The induced representation $\chi_1 \times \chi_2 \rtimes \sigma$ is irreducible if and only if $\chi_1 \neq \nu^{\pm 1}, \chi_2 \neq \nu^{\pm 1}$ and $\chi_1 \neq \nu^{\pm 1}\chi_2^{\pm 1}$. Here, ν is the unramified character of $\mathbb{Q}_p^\times$ defined by $\nu(a) = |a|$. Groups VII, VIII, and IX contain representations supported in Q, i.e., they are constituents of induced representations of the form $\chi \rtimes \pi$, where π is a supercuspidal representation of $GL_2(\mathbb{Q}_p)$. Finally, groups X and XI contain representations supported in P, i.e., they are constituents of induced representations of the form $\pi \rtimes \sigma$, where π is a supercuspidal representation of $GL_2(\mathbb{Q}_p)$. All of this data (and more) is included in Table 7.1 of irreducible, non-supercuspidal representations of G_p. In Table 7.2, we have the conditions for unitarity.

Exercise 7.2 *Let $(\pi, V) = \eta^{-1} \times \eta$ be an induced representation of $GL_2(\mathbb{Q}_p)$. Let $\tau = |\det|^{-1/2}\pi$. Show that $\tau \rtimes \nu^{1/2} \simeq \nu^{-1/2}\eta \times \nu^{-1/2}\eta^{-1} \rtimes \nu^{1/2}$.*

Table 7.1 These are the non-supercuspidal representations of G_p

		Constituent of	Representation	Tempered	L^2	g
I		$\chi_1 \times \chi_2 \rtimes \sigma$ (irreducible)		χ_i, σ unit.		•
II	a	$\nu^{1/2}\chi \times \nu^{-1/2}\chi \rtimes \sigma$	$\chi St_{GL(2)} \rtimes \sigma$	χ, σ unit.		•
	b	$(\chi^2 \neq \nu^{\pm 1}, \chi \neq \nu^{\pm 3/2})$	$\chi \mathbf{1}_{GL(2)} \rtimes \sigma$			
III	a	$\chi \times \nu \rtimes \nu^{-1/2}\sigma$	$\chi \rtimes \sigma St_{GSp(2)}$	π, σ unit.		•
	b	$(\chi \notin \{1, \nu^{\pm 2}\})$	$\chi \rtimes \sigma \mathbf{1}_{GSp(2)}$			
IV	a		$\sigma St_{GSp(4)}$	σ unit.	•	•
	b		$L(\nu^2, \nu^{-1}\sigma St_{GSp(2)})$			
	c	$\nu^2 \times \nu \rtimes \nu^{-3/2}\sigma$	$L(\nu^{3/2}St_{GL(2)}, \nu^{-3/2}\sigma)$			
	d		$\sigma \mathbf{1}_{GSp(4)}$			
V	a		$\delta([\xi, \nu\xi], \nu^{-1/2}\sigma)$	σ unit.	•	•
	b	$\nu\xi \times \xi \rtimes \nu^{-1/2}\sigma$	$L(\nu^{1/2}\xi St_{GL(2)}, \nu^{-1/2}\sigma)$			
	c	$(\xi^2 = 1, \xi \neq 1)$	$L(\nu^{1/2}\xi St_{GL(2)}, \xi\nu^{-1/2}\sigma)$			
	d		$L(\nu\xi, \xi \rtimes \nu^{-1/2}\sigma)$			
VI	a		$\tau(S, \nu^{-1/2}\sigma)$	σ unit.		•
	b		$\tau(T, \nu^{-1/2}\sigma)$	σ unit.		
	c	$\nu \times 1_{F^\times} \rtimes \nu^{-1/2}\sigma$	$L(\nu^{1/2}St_{GL(2)}, \nu^{-1/2}\sigma)$			
	d		$L(\nu, 1_{F^\times} \rtimes \nu^{-1/2}\sigma)$			
VII		$\chi \rtimes \pi$ (irreducible)		χ, π unit.		•
VIII	a		$\tau(S, \pi)$	π unit.		•
	b	$1_{F^\times} \rtimes \pi$	$\tau(T, \pi)$	π unit.		
IX	a	$\nu\xi \rtimes \nu^{-1/2}\pi$	$\delta(\nu\xi, \nu^{-1/2}\pi)$	π unit.	•	•
	b	$(\xi \neq 1, \xi\pi = \pi)$	$L(\nu\xi, \nu^{-1/2}\pi)$			
X		$\pi \rtimes \sigma$ (irreducible)		π, σ unit.		•
XI	a	$\nu^{1/2}\pi \rtimes \nu^{-1/2}\sigma$	$\delta(\nu^{1/2}\pi, \nu^{-1/2}\sigma)$	π, σ unit.	•	•
	b	$(\omega_\pi = 1)$	$L(\nu^{1/2}\pi, \nu^{-1/2}\sigma)$			

7.2   Generic Representations

Fix a nontrivial additive character ψ of $\mathbb{Q}_p$. Every other such character is given by $x \mapsto \psi(cx)$ for $c \in \mathbb{Q}_p^\times$. Let c_1, c_2 be two elements of $\mathbb{Q}_p^\times$. We obtain a character of the unipotent radical N of the Borel subgroup B as follows.

$$\psi_{c_1,c_2}\left(\begin{bmatrix} 1 & y & * & \\ x & 1 & * & * \\ & & 1 & -x \\ & & & 1 \end{bmatrix}\right) = \psi(c_1 x + c_2 y).$$

Table 7.2 Conditions for unitarity

		Representation	Conditions for unitarity
I		$\chi_1 \times \chi_2 \rtimes \sigma$ (irreducible)	$e(\chi_1) = e(\chi_2) = e(\sigma) = 0$
			$\chi_1 = \nu^\beta \chi$, $\chi_2 = \nu^\beta \chi^{-1}$, $e(\sigma) = -\beta$, $e(\chi) = 0$, $\chi^2 \neq 1$, $0 < \beta < 1/2$
			$\chi_1 = \nu^\beta$, $e(\chi_2) = 0$, $e(\sigma) = -\beta/2$, $\chi_2 \neq 1$, $0 < \beta < 1$
			$\chi_1 = \nu^{\beta_1}\chi$, $\chi_2 = \nu^{\beta_2}\chi$, $e(\sigma) = (-\beta_1 - \beta_2)/2$, $\chi^2 = 1$, $0 \leq \beta_2 \leq \beta_1$, $0 < \beta_1 < 1$, $\beta_1 + \beta_2 < 1$
II	a	$\chi \mathrm{St}_{\mathrm{GL}(2)} \rtimes \sigma$	$e(\sigma) = e(\chi) = 0$
			$\chi = \xi\nu^\beta$, $e(\sigma) = -\beta$, $\xi^2 = 1$, $0 < \beta < 1/2$
	b	$\chi \mathbf{1}_{\mathrm{GL}(2)} \rtimes \sigma$	$e(\sigma) = e(\chi) = 0$
			$\chi = \xi\nu^\beta$, $e(\sigma) = -\beta$, $\xi^2 = 1$, $0 < \beta < 1/2$
III	a	$\chi \rtimes \sigma\mathrm{St}_{\mathrm{GSp}(2)}$	$e(\sigma) = e(\chi) = 0$
	b	$\chi \rtimes \sigma\mathbf{1}_{\mathrm{GSp}(2)}$	$e(\sigma) = e(\chi) = 0$
IV	a	$\sigma\mathrm{St}_{\mathrm{GSp}(4)}$	$e(\sigma) = 0$
	b	$L(\nu^2, \nu^{-1}\sigma\mathrm{St}_{\mathrm{GSp}(2)})$	Never unitary
	c	$L(\nu^{3/2}\mathrm{St}_{\mathrm{GSp}(2)}, \nu^{-3/2}\sigma)$	Never unitary
	d	$\sigma\mathbf{1}_{\mathrm{GSp}(4)}$	$e(\sigma) = 0$
V	a	$\delta([\xi, \nu\xi], \nu^{-1/2}\sigma)$	$e(\sigma) = 0$
	b	$L(\nu^{1/2}\xi\mathrm{St}_{\mathrm{GL}(2)}, \nu^{-1/2}\sigma)$	$e(\sigma) = 0$
	c	$L(\nu^{1/2}\xi\mathrm{St}_{\mathrm{GL}(2)}, \xi\nu^{-1/2}\sigma)$	$e(\sigma) = 0$
	d	$L(\nu\xi, \xi \rtimes \nu^{-1/2}\sigma)$	$e(\sigma) = 0$
VI	a	$\tau(S, \nu^{-1/2}\sigma)$	$e(\sigma) = 0$
	b	$\tau(T, \nu^{-1/2}\sigma)$	$e(\sigma) = 0$
	c	$L(\nu^{1/2}\mathrm{St}_{\mathrm{GL}(2)}, \nu^{-1/2}\sigma)$	$e(\sigma) = 0$
	d	$L(\nu, 1_{F^\times} \rtimes \nu^{-1/2}\sigma)$	$e(\sigma) = 0$
VII		$\chi \rtimes \pi$ (irreducible)	$e(\chi) = e(\pi) = 0$
			$\chi = \nu^\beta\xi$, $\pi = \nu^{-\beta/2}\rho$, $0 < \beta < 1$, $\xi^2 = 1$, $\xi \neq 1$, $e(\rho) = 0$, $\xi\rho = \rho$
VIII	a	$\tau(S, \pi)$	$e(\pi) = 0$
	b	$\tau(T, \pi)$	$e(\pi) = 0$
IX	a	$\delta(\nu\xi, \nu^{-1/2}\pi)$	$e(\pi) = 0$
	b	$L(\nu\xi, \nu^{-1/2}\pi)$	$e(\pi) = 0$
X		$\pi \rtimes \sigma$ (irreducible)	$e(\sigma) = e(\pi) = 0$
			$\pi = \nu^\beta\rho$, $e(\sigma) = -\beta$, $0 < \beta < 1/2$, $\omega_\rho = 1$
XI	a	$\delta(\nu^{1/2}\pi, \nu^{-1/2}\sigma)$	$e(\sigma) = e(\pi) = 0$
	b	$L(\nu^{1/2}\pi, \nu^{-1/2}\sigma)$	$e(\sigma) = e(\pi) = 0$

An irreducible admissible representation π of G_p is called *generic* if there is a nonzero linear functional ℓ on V_π satisfying

$$\ell\left(\pi\left(\begin{bmatrix} 1 & y & * & \\ x & 1 & * & * \\ & & 1 & -x \\ & & & 1 \end{bmatrix}\right)v\right) = \psi(c_1 x + c_2 y)\ell(v), \qquad \text{for all } v \in V_\pi,$$

i.e., $\mathrm{Hom}_{N(\mathbb{Q}_p)}(\pi, \psi_{c_1, c_2}) \neq 0$. Such a nonzero functional is called a *Whittaker functional*.

Exercise 7.3 *Let π be generic and let $0 \neq \ell \in \mathrm{Hom}_{N(\mathbb{Q}_p)}(\pi, \psi_{c_1, c_2})$. For every $v \in V_\pi$, define a $\mathbb{C}$-valued function W_v on G_p by*

$$W_v(g) := \ell(\pi(g)v).$$

Show that W_v satisfies

$$W_v\left(\begin{bmatrix} 1 & y & * & \\ x & 1 & * & * \\ & & 1 & -x \\ & & & 1 \end{bmatrix} g\right) = \psi(c_1 x + c_2 y) W_v(g), \qquad \text{for all } g \in G_p. \qquad (7.1)$$

The above exercise tells us that a generic representation has a *Whittaker model*, i.e., it is isomorphic to a space of $\mathbb{C}$-valued functions on G_p which satisfy (7.1). We will denote the Whittaker model by $\mathcal{W}(\pi, \psi_{c_1, c_2})$.

Exercise 7.4 *Suppose π is given by its Whittaker model $\mathcal{W}(\pi, \psi_{c_1, c_2})$. Give a formula for a Whittaker functional on π.*

It has been shown in [83] that if π has a Whittaker model, then it is unique. The independence of existence on c_1, c_2 is shown in the next exercise.

Exercise 7.5 *Suppose that $\mathrm{Hom}_{N(\mathbb{Q}_p)}(\pi, \psi_{1,1}) \neq 0$. Then show that, for any $c_1, c_2 \in \mathbb{Q}_p^\times$, we have $\mathrm{Hom}_{N(\mathbb{Q}_p)}(\pi, \psi_{c_1, c_2}) \neq 0$.*

The last column in Table 7.1 states exactly which non-supercuspidal representation is generic. Note that each group has exactly one which is generic.

7.3 Iwahori-Spherical Representations

In the previous chapter, we have seen that if χ_1, χ_2, σ are unramified, then $\chi_1 \times \chi_2 \rtimes \sigma$ has a unique spherical constituent. This means that, out of all the representations of Types II–VI, there is only one in the group that is spherical. It turns out that all the other representations do have vectors invariant under some special compact subgroups called the parahoric subgroups. Define the Iwahori subgroup I of G_p by

$$I = K_p \cap \begin{bmatrix} \mathbb{Z}_p & p\mathbb{Z}_p & \mathbb{Z}_p & \mathbb{Z}_p \\ \mathbb{Z}_p & \mathbb{Z}_p & \mathbb{Z}_p & \mathbb{Z}_p \\ p\mathbb{Z}_p & p\mathbb{Z}_p & \mathbb{Z}_p & \mathbb{Z}_p \\ p\mathbb{Z}_p & p\mathbb{Z}_p & p\mathbb{Z}_p & \mathbb{Z}_p \end{bmatrix}. \tag{7.2}$$

Recall that $K_p = \mathrm{GSp}_4(\mathbb{Z}_p)$. The Atkin–Lehner element

$$\eta_1 = \begin{bmatrix} & & & -1 \\ & & 1 & \\ & p & & \\ -p & & & \end{bmatrix}$$

normalizes I. There are 4 other *parahoric* subgroups of G_p. They are given by

$$K_p, \quad P_1 = K_p \cap \begin{bmatrix} \mathbb{Z}_p & \mathbb{Z}_p & \mathbb{Z}_p & \mathbb{Z}_p \\ \mathbb{Z}_p & \mathbb{Z}_p & \mathbb{Z}_p & \mathbb{Z}_p \\ p\mathbb{Z}_p & p\mathbb{Z}_p & \mathbb{Z}_p & \mathbb{Z}_p \\ p\mathbb{Z}_p & p\mathbb{Z}_p & \mathbb{Z}_p & \mathbb{Z}_p \end{bmatrix}, \quad P_2 = K_p \cap \begin{bmatrix} \mathbb{Z}_p & p\mathbb{Z}_p & \mathbb{Z}_p & \mathbb{Z}_p \\ \mathbb{Z}_p & \mathbb{Z}_p & \mathbb{Z}_p & \mathbb{Z}_p \\ \mathbb{Z}_p & p\mathbb{Z}_p & \mathbb{Z}_p & \mathbb{Z}_p \\ p\mathbb{Z}_p & p\mathbb{Z}_p & p\mathbb{Z}_p & \mathbb{Z}_p \end{bmatrix},$$

$$P_{0,2} = \{k \in G_p \cap \begin{bmatrix} \mathbb{Z}_p & p\mathbb{Z}_p & \mathbb{Z}_p & \mathbb{Z}_p \\ \mathbb{Z}_p & \mathbb{Z}_p & \mathbb{Z}_p & p^{-1}\mathbb{Z}_p \\ \mathbb{Z}_p & p\mathbb{Z}_p & \mathbb{Z}_p & \mathbb{Z}_p \\ p\mathbb{Z}_p & p\mathbb{Z}_p & p\mathbb{Z}_p & \mathbb{Z}_p \end{bmatrix} : \det(k) \in \mathbb{Z}_p^\times\}.$$

P_1 is the Siegel congruence subgroup, P_2 is the Klingen congruence subgroup, and $P_{0,2}$ is the paramodular group, which is a maximal compact subgroup of G_p which is not conjugate to K_p.

Exercise 7.6 *Let*

$$s_0 = \begin{bmatrix} 1 & & & \\ & & -p^{-1} & \\ & 1 & & \\ p & & & \end{bmatrix}, \quad s_1 = \begin{bmatrix} & 1 & & \\ 1 & & & \\ & & & 1 \\ & & 1 & \end{bmatrix}, \quad s_2 = \begin{bmatrix} 1 & & & \\ & & 1 & \\ & -1 & & \\ & & & 1 \end{bmatrix}.$$

For a subset S of $\{s_0, s_1, s_2\}$, define $P_S := \sqcup_{s \in \langle S \rangle} \mathrm{I} s \mathrm{I}$. Show that $P_{s_1, s_2} = K_p$, $P_{s_0, s_2} = P_{0,2}$, $P_{s_1} = P_1$, $P_{s_2} = P_2$. Here, $\langle S \rangle$ is the group generated by the elements of S.

An irreducible admissible representation π of G_p is called *Iwahori-spherical* if π contains a vector that is invariant under I. It is known that any Iwahori-spherical representation is obtained as a subquotient of $\chi_1 \times \chi_2 \rtimes \sigma$ for unramified characters χ_1, χ_2, σ. Schmidt [92] has obtained the dimensions of vectors invariant under the various parahoric subgroups, and these are given in Table 7.3 below. Since η_1 normalizes I, it acts on the I-invariant vectors in π. This is a finite dimensional vector space, and you can diagonalize η_1 on this space. Note that, since $\eta_1^2 = \eta_1$, the eigenvalues are ± 1.

Table 7.3 The Iwahori-spherical representations of G_p, and the dimensions of their spaces of fixed vectors under the parahoric subgroups

		π	K_p	$P_{0,2}$	P_2	P_1	I
I		$\chi_1 \times \chi_2 \rtimes \sigma$ (irreducible)	1	2	4	4	8
				+−		++	++++
						−−	−−−−
II	a	$\chi St_{GL(2)} \rtimes \sigma$	0	1	2	1	4
				−		−	+−−−
	b	$\chi 1_{GL(2)} \rtimes \sigma$	1	1	2	3	4
				+		++−	+++−
III	a	$\chi \rtimes \sigma St_{GSp(2)}$	0	0	1	2	4
						+−	++−−
	b	$\chi \rtimes \sigma 1_{GSp(2)}$	1	2	3	2	4
				+−		+−	++−−
IV	a	$\sigma St_{GSp(4)}$	0	0	0	0	1
							−
	b	$L(\nu^2, \nu^{-1}\sigma St_{GSp(2)})$	0	0	1	2	3
						+−	++−
	c	$L(\nu^{3/2}St_{GL(2)}, \nu^{-3/2}\sigma)$	0	1	2	1	3
				−		−	+−−
	d	$\sigma 1_{GSp(4)}$	1	1	1	1	1
				+		+	+
V	a	$\delta([\xi, \nu\xi], \nu^{-1/2}\sigma)$	0	0	1	0	2
							+−
	b	$L(\nu^{1/2}\xi St_{GL(2)}, \nu^{-1/2}\sigma)$	0	1	1	1	2
				+		+	++
	c	$L(\nu^{1/2}\xi St_{GL(2)}, \xi\nu^{-1/2}\sigma)$	0	1	1	1	2
				−		−	−−
	d	$L(\nu\xi, \xi \rtimes \nu^{-1/2}\sigma)$	1	0	1	2	2
						+−	+−
VI	a	$\tau(S, \nu^{-1/2}\sigma)$	0	0	1	1	3
						−	+−−
	b	$\tau(T, \nu^{-1/2}\sigma)$	0	0	0	1	1
						+	+
	c	$L(\nu^{1/2}St_{GL(2)}, \nu^{-1/2}\sigma)$	0	1	1	0	1
				−			−
	d	$L(\nu, 1_{F^\times} \rtimes \nu^{-1/2}\sigma)$	1	1	2	2	3
				+		+−	++−

Schmidt used the local theory of Iwahori-spherical representations of G_p to formulate a new-forms theory for Siegel cusp forms with square-free level. Let $N > 1$ be a square-free integer. The space of old forms of weight k and level N, denoted by $S_k(B(N))^{old}$, is sum of the spaces

$$S_k(B(N_1) \cap \eta_{N_2} Q(N_2)\eta_{N_2}^{-1}) + S_k(B(N_1) \cap P(N_2)) + S_k(B(N_1) \cap Q(N_2)),$$

where N_1, N_2 runs through all positive integers such that $N_1 N_2 = N$, $N_2 > 1$. The orthogonal complement of $S_k(B(N))^{old}$ in $S_k(B(N))$ is defined to be the space of new-forms $S_k(B(N))^{new}$.

If $F \in S_k(B(N))^{new}$ is a Hecke eigenform for all $p \nmid N$, one can construct Φ_F on $G(\mathbb{A})$ which generates a representation π_F. Let $\pi_F \cong \otimes \pi_p$. Then, it can be shown that for primes $p|N$, π_p is a Type IVa representation—the Steinberg representation for G_p.

7.4 Paramodular Theory

The local analogue of the paramodular congruence subgroup is given by

$$K(n) := \{k \in G_p \cap \begin{bmatrix} \mathbb{Z}_p & p^n\mathbb{Z}_p & \mathbb{Z}_p & \mathbb{Z}_p \\ \mathbb{Z}_p & \mathbb{Z}_p & \mathbb{Z}_p & p^{-n}\mathbb{Z}_p \\ \mathbb{Z}_p & p^n\mathbb{Z}_p & \mathbb{Z}_p & \mathbb{Z}_p \\ p^n\mathbb{Z}_p & p^n\mathbb{Z}_p & p^n\mathbb{Z}_p & \mathbb{Z}_p \end{bmatrix} : \det(k) \in \mathbb{Z}_p^\times\}, \qquad n \geq 0. \tag{7.3}$$

Note that $K(0) = K_p$. Also, we do *not* have $K(n+1) \subset K(n)$.

Exercise 7.7 *The Atkin–Lehner element is defined by*

$$\eta_n := \begin{bmatrix} & & & -1 \\ & & 1 & \\ & p^n & & \\ -p^n & & & \end{bmatrix}$$

Show that $K(n)$ is normalized by η_n. Hence, show that if there is a vector v in a representation π of G_p that is invariant under $K(n)$, then so is $\pi(\eta_n)v$.

Given an irreducible admissible representation (π, V) of G_p, define

$$V(n) := \{v \in V : \pi(k)v = v, \text{ for all } k \in K(n)\}.$$

If there is a $n \geq 0$ such that $V(n) \neq 0$, then we say that π is a *paramodular* representation. If π is paramodular, then define the paramodular level of π to be the smallest nonnegative integer N_π such that $V(N_\pi) \neq 0$. For example, if π is spherical then clearly π is paramodular and has paramodular level 0.

An important tool to understand the paramodular theory is the paramodular Hecke algebra $\mathcal{H}(G_p, K(n))$ of compactly supported functions on G_p that are left and right $K(n)$-invariant.

Exercise 7.8 *Show that, if $v \in V(n)$ then $Tv \in V(n)$ for all $T \in \mathcal{H}(G_p, K(n))$.*

In particular, there are two Hecke operators that play a significant role

$$
T_{0,1} = \mathrm{char}\!\left(K(n) \begin{bmatrix} p & & & \\ & p & & \\ & & 1 & \\ & & & 1 \end{bmatrix} K(n) \right), \quad
T_{1,0} = \mathrm{char}\!\left(K(n) \begin{bmatrix} p & & & \\ & p^2 & & \\ & & p & \\ & & & 1 \end{bmatrix} K(n) \right).
$$

For $n = 0$ we have the following single coset decompositions (see [81, Pg 187–188]).

$$
K_p \begin{bmatrix} p & & & \\ & p & & \\ & & 1 & \\ & & & 1 \end{bmatrix} K_p = \bigsqcup_{x,y,z \in \mathbb{Z}_p/p\mathbb{Z}_p} \begin{bmatrix} 1 & x & y & \\ & 1 & y & z \\ & & 1 & \\ & & & 1 \end{bmatrix} \begin{bmatrix} p & & & \\ & p & & \\ & & 1 & \\ & & & 1 \end{bmatrix} K_p
$$

$$
\sqcup \bigsqcup_{x \in \mathbb{Z}_p/p\mathbb{Z}_p} \begin{bmatrix} 1 & x & & \\ & 1 & & \\ & & 1 & \\ & & & 1 \end{bmatrix} \begin{bmatrix} p & & & \\ & 1 & & \\ & & 1 & \\ & & & p \end{bmatrix} K_p \sqcup \begin{bmatrix} 1 & & & \\ & 1 & & \\ & & p & \\ & & & p \end{bmatrix} K_p
$$

$$
\sqcup \bigsqcup_{x,z \in \mathbb{Z}_p/p\mathbb{Z}_p} \begin{bmatrix} 1 & & & \\ x & 1 & & \\ & & 1 & -x \\ & & & 1 \end{bmatrix} \begin{bmatrix} 1 & & & \\ & 1 & z & \\ & & 1 & \\ & & & 1 \end{bmatrix} \begin{bmatrix} 1 & & & \\ & p & & \\ & & p & \\ & & & 1 \end{bmatrix} K_p
$$

$$
K_p \begin{bmatrix} p & & & \\ & p^2 & & \\ & & 1 & \\ & & & p \end{bmatrix} K_p = \bigsqcup_{\substack{z \in \mathbb{Z}_p/p^2\mathbb{Z}_p \\ x,y \in \mathbb{Z}_p/p\mathbb{Z}_p}} \begin{bmatrix} 1 & & & \\ x & 1 & & \\ & & 1 & -x \\ & & & 1 \end{bmatrix} \begin{bmatrix} 1 & y & & \\ & 1 & y & z \\ & & 1 & \\ & & & 1 \end{bmatrix} \begin{bmatrix} p & & & \\ & p^2 & & \\ & & p & \\ & & & 1 \end{bmatrix} K_p
$$

$$
\sqcup \bigsqcup_{\substack{d \in \mathbb{Z}_p/p^2\mathbb{Z}_p \\ c \in \mathbb{Z}_p/p\mathbb{Z}_p}} \begin{bmatrix} 1 & d & c & \\ & 1 & c & \\ & & 1 & \\ & & & 1 \end{bmatrix} \begin{bmatrix} p^2 & & & \\ & p & & \\ & & 1 & \\ & & & p \end{bmatrix} K_p \sqcup \begin{bmatrix} p & & & \\ & 1 & & \\ & & p & \\ & & & p^2 \end{bmatrix} K_p
$$

$$
\sqcup \bigsqcup_{x \in \mathbb{Z}_p/p\mathbb{Z}_p} \begin{bmatrix} 1 & & & \\ x & 1 & & \\ & & 1 & -x \\ & & & 1 \end{bmatrix} \begin{bmatrix} p & & & \\ & p^2 & & \\ & & p & \\ & & & p \end{bmatrix} K_p
$$

$$
\sqcup \bigsqcup_{d \in (\mathbb{Z}_p/p\mathbb{Z}_p)^\times} \begin{bmatrix} 1 & d/p & & \\ & 1 & & \\ & & 1 & \\ & & & 1 \end{bmatrix} \begin{bmatrix} p & & & \\ & p & & \\ & & p & \\ & & & p \end{bmatrix} K_p
$$

$$\bigsqcup_{\substack{u\in(\mathbb{Z}_p/p\mathbb{Z}_p)^\times \\ \lambda\in\mathbb{Z}_p/p\mathbb{Z}_p}} \begin{bmatrix} 1 & \lambda^2 u/p & \lambda u/p \\ & 1 & \lambda u/p & u/p \\ & & 1 & \\ & & & 1 \end{bmatrix} \begin{bmatrix} p & & & \\ & p & & \\ & & p & \\ & & & p \end{bmatrix} K_p.$$

Exercise 7.9 *Let* χ_1, χ_2, σ *be unramified characters with* $\chi_1\chi_2\sigma^2 = 1$ *and let* $\pi = \chi_1 \times \chi_2 \rtimes \sigma$ *be given by its induced model. Let* f_0 *be the essentially unique nonzero vector in* V_π *which is invariant under* K_p. *Let* λ, μ *be the Hecke eigenvalues defined by* $T_{0,1} f_0 = \lambda f_0$ *and* $T_{1,0} f_0 = \mu f_0$. *Use the coset decomposition above to show that*

$$\lambda = p^{3/2}\sigma(p)(1 + \chi_1(p))(1 + \chi_2(p)),$$
$$\mu = p^2(\chi_1(p) + \chi_1^{-1}(p) + \chi_2(p) + \chi_2^{-1}(p) + 1 - p^{-2}).$$

The following new forms theory has been obtained by Roberts and Schmidt in [81].

Theorem 7.10 (Roberts, Schmidt [81]) *Let* (π, V) *be an irreducible admissible representation of* G_p *with trivial central character.*

(i) If π *is paramodular, and* N_π *is the paramodular level, then* $\dim(V(N_\pi)) = 1$.
(ii) Assume that π *is generic and given by the Whittaker model* $\mathcal{W}(\pi, \psi_{c_1,c_2})$, *with* $c_1, c_2 \in \mathbb{Z}_p^\times$, *and conductor of* ψ *equal to* $\mathbb{Z}_p$. *Then* π *is paramodular. For any* $W \in \mathcal{W}(\pi, \psi_{c_1,c_2})$, *define the zeta integral*

$$Z(s, W) := \int_{\mathbb{Q}_p^\times} \int_{\mathbb{Q}_p} W(\begin{bmatrix} a & & & \\ & a & & \\ x & & 1 & \\ & & & 1 \end{bmatrix})|a|^{s-\frac{3}{2}} dx d^\times a.$$

Then, there is a $W_\pi \in V(N_\pi)$ *such that* $Z(s, W_\pi) = L(s, \pi)$. *Here, the local L-factors* $L(s, \pi)$ *are defined in [34] or [102].*
(iii) Assume that π *is generic, and let* λ_π *and* μ_π *be the Hecke eigenvalues of* $T_{0,1}$ *and* $T_{1,0}$ *acting on a nonzero element* W *of* $V(N_\pi)$.

(a) Let $N_\pi = 0$, *so that* π *is unramified. Then*

$$L(s, \pi)^{-1} = 1 - p^{-\frac{3}{2}}\lambda_\pi p^{-s} + (p^{-2}\mu_\pi + 1 + p^{-2})p^{-2s} - p^{-\frac{3}{2}}\lambda_\pi p^{-3s} + p^{-4s}.$$

(b) Let $N_\pi = 1$ *and let* $\pi(\eta_1)W = \varepsilon_\pi W$, *where* $\varepsilon_\pi \in \{1, -1\}$ *is the Atkin–Lehner eigenvalue of* W. *Then*

$$L(s, \pi)^{-1} = 1 - p^{-\frac{3}{2}}(\lambda_\pi + \mu_\pi)p^{-s} + (p^{-2}\mu_\pi + 1)p^{-2s} + \varepsilon_\pi p^{-\frac{1}{2}}p^{-3s}.$$

(c) If $N_\pi \geq 2$, *then*

$$L(s, \pi)^{-1} = 1 - p^{-\frac{3}{2}}\lambda_\pi p^{-s} + (p^{-2}\mu_\pi + 1)p^{-2s}.$$

Table 7.4 Hecke eigenvalues (Table A.14 of [81])

	Inducing data	N_π	λ_π	μ_π
I	σ, χ_1, χ_2 unr.	0	$q^{3/2}\sigma(p)\big(1 + \chi_1(p)$ $+\chi_2(p) + \chi_1(p)\chi_2(p)\big)$	(A)
	σ unr., χ_1, χ_2 ram.	$a(\chi_1) +$ $a(\chi_2)$	$p^{3/2}(\sigma(p) + \sigma(p)^{-1})$	0
	σ ram., $\sigma\chi_i$ unr.	$2a(\sigma)$	$p^{3/2}((\chi_1\sigma)(p) +$ $(\chi_2\sigma)(p))$	0
	σ ram., $\sigma\chi_i$ ram.	$2a(\chi_1\sigma) +$ $2a(\sigma)$	0	$-p^2$
IIa			$p^{3/2}(\sigma(p) + \sigma(p)^{-1})$	$p^{3/2}(\chi(p)$
	σ, χ unr.	1	$+(p + 1)(\sigma\chi)(p)$	$+\chi(p)^{-1})$
	σ, χ ram., $\chi\sigma$ unr.	$2a(\sigma) + 1$	$p(\chi\sigma)(p)$	$-p^2$
	σ unr., $\chi\sigma$ ram.	$2a(\chi)$	$p^{3/2}(\sigma(p) + \sigma(p)^{-1})$	0
	σ ram., $\chi\sigma$ ram.	$2a(\sigma) +$ $2a(\chi\sigma)$	0	$-p^2$
IIb			$p^{3/2}(\sigma(p) + \sigma(p)^{-1})$	
	σ, χ unr.	0	$+p(p + 1)(\sigma\chi)(p)$	(B)
	σ, χ ram., $\chi\sigma$ unr.	$2a(\sigma)$	$p(p + 1)(\sigma\chi)(p)$	0
IIIa	$\chi\sigma$ ram.	—	—	—
	σ unr.	2	$p(\sigma(p) + \sigma(p)^{-1})$	-p(p-1)
	σ ram.	$4a(\sigma)$	0	$-p^2$
IIIb	σ unr.	0	$p(p + 1)\sigma(p)(1 +$ $\chi(p))$	(C)
	σ ram.	—	—	—
IVa	σ unr.	3	$\sigma(p)$	$-p^2$
	σ ram.	$4a(\sigma)$	0	$-p^2$
IVb	σ unr.	2	$\sigma(p)(1 + p^2)$	$-p(p - 1)$
	σ ram.	—	—	—
IVc	σ unr.	1	$\sigma(p)(p^3 + p + 2)$	$p^3 + 1$
	σ ram.	—	—	—
IVd	σ unr.	0	$\sigma(p)(p^3 + p^2 + p + 1)$	$p(p^3 + p^2 +$ $p + 1)$
	σ ram.	—	—	—
Va	ξ, σ unr.	2	0	$-p^2 - p$
	σ unr., ξ ram.	$2a(\xi) + 1$	$\sigma(p)p$	$-p^2$
	σ ram., $\sigma\xi$ unr.	$2a(\sigma) + 1$	$-\sigma(p)p$	$-p^2$
	$\sigma, \sigma\xi$ ram.	$2a(\xi\sigma) +$ $2a(\sigma)$	0	$-p^2$

(continued)

Table 7.4 (continued)

	inducing data	N_π	λ_π	μ_π
Vb	ξ, σ unr.	1	$\sigma(p)(p^2 - 1)$	$-p^2 - p$
	σ unr., ξ ram.	$2a(\xi)$	$\sigma(p)p(p + 1)$	0
	σ ram., $\sigma\xi$ unr.	—	—	—
	$\sigma, \sigma\xi$ ram.	—	—	—
Vc	inducing data	N_π	λ_π	μ_π
	ξ, σ unr.	1	$-\sigma(p)(p^2 - 1)$	$-p^2 - p$
	σ unr., ξ ram.	—	—	—
	σ ram., $\sigma\xi$ unr.	$2a(\sigma)$	$-\sigma(p)p(p + 1)$	0
	$\sigma, \sigma\xi$ ram.	—	—	—
Vd	ξ, σ unr.	0	0	$-(p^3 + p^2 + p + 1)$
	ξ or σ ram.	—	—	—
VIa	σ unr.	2	$2p\sigma(p)$	$-p(p - 1)$
	σ ram.	$4a(\sigma)$	0	$-p^2$
VIb	σ unr.	—	—	—
	σ ram.	—	—	—
VIc	σ unr.	1	$\sigma(p)(p + 1)^2$	$p(p + 1)$
	σ ram.	—	—	—
VId	σ unr.	0	$2p(p + 1)\sigma(p)$	$(p + 1)(p^2 + 2p - 1)$
	σ ram.	—	—	—
VII		$2a(\pi)$	0	$-p^2$
VIIIa		$2a(\pi)$	0	$-p^2$
VIIIb		—	—	—
IXa		$2a(\pi) + 1$	0	$-p^2$
IXb		—	—	—
X	σ unr.	$a(\sigma\pi)$	$p^{3/2}(\sigma(p) + \sigma(p)^{-1})$	0
	σ ram.	$a(\sigma\pi) + 2a(\sigma)$	0	$-p^2$
XIa	σ unr.	$a(\sigma\pi) + 1$	$p\sigma(p)$	$-p^2$
	σ ram.	$a(\sigma\pi) + 2a(\sigma)$	0	$-p^2$
XIb	σ unr.	$a(\sigma\pi)$	$p(p + 1)\sigma(p)$	0
	σ ram.	—	—	—
super-cuspidal	generic	≥ 2	0	$-p^2$
	non-generic	—	—	—

Table 7.5 The following table gives the degree 4 L-factors of all the non-supercuspidal representations of G_p (Table A.8 of [81])

		Representation	$L(s,\pi)$
I		$\chi_1 \times \chi_2 \rtimes \sigma$ (irreducible)	$L(s,\chi_1\chi_2\sigma)L(s,\sigma)L(s,\chi_1\sigma)L(s,\chi_2\sigma)$
II	a	$\chi St_{GL(2)} \rtimes \sigma$	$L(s,\chi^2\sigma)L(s,\sigma)L(s,\nu^{1/2}\chi\sigma)$
	b	$\chi 1_{GL(2)} \rtimes \sigma$	$L(s,\chi^2\sigma)L(s,\sigma)L(s,\nu^{1/2}\chi\sigma)L(s,\nu^{-1/2}\chi\sigma)$
III	a	$\chi \rtimes \sigma St_{GSp(2)}$	$L(s,\nu^{1/2}\chi\sigma)L(s,\nu^{1/2}\sigma)$
	b	$\chi \rtimes \sigma 1_{GSp(2)}$	$L(s,\nu^{1/2}\chi\sigma)L(s,\nu^{1/2}\sigma)L(s,\nu^{-1/2}\chi\sigma)L(s,\nu^{-1/2}\sigma)$
IV	a	$\sigma St_{GSp(4)}$	$L(s,\nu^{3/2}\sigma)$
	b	$L(\nu^2,\nu^{-1}\sigma St_{GSp(2)})$	$L(s,\nu^{3/2}\sigma)L(s,\nu^{-1/2}\sigma)$
	c	$L(\nu^{3/2}St_{GL(2)},\nu^{-3/2}\sigma)$	$L(s,\nu^{3/2}\sigma)L(s,\nu^{1/2}\sigma)L(s,\nu^{-3/2}\sigma)$
	d	$\sigma 1_{GSp(4)}$	$L(s,\nu^{3/2}\sigma)L(s,\nu^{1/2}\sigma)L(s,\nu^{-1/2}\sigma)L(s,\nu^{-3/2}\sigma)$
V	a	$\delta([\xi,\nu\xi],\nu^{-1/2}\sigma)$	$L(s,\nu^{1/2}\sigma)L(s,\nu^{1/2}\xi\sigma)$
	b	$L(\nu^{1/2}\xi St_{GL(2)},\nu^{-1/2}\sigma)$	$L(s,\nu^{1/2}\sigma)L(s,\nu^{1/2}\xi\sigma)L(s,\nu^{-1/2}\sigma)$
	c	$L(\nu^{1/2}\xi St_{GL(2)},\xi\nu^{-1/2}\sigma)$	$L(s,\nu^{1/2}\sigma)L(s,\nu^{1/2}\xi\sigma)L(s,\nu^{-1/2}\xi\sigma)$
	d	$L(\nu\xi,\xi \rtimes \nu^{-1/2}\sigma)$	$L(s,\nu^{1/2}\sigma)L(s,\nu^{1/2}\xi\sigma)L(s,\nu^{-1/2}\sigma)L(s,\nu^{-1/2}\xi\sigma)$
VI	a	$\tau(S,\nu^{-1/2}\sigma)$	$L(s,\nu^{1/2}\sigma)^2$
	b	$\tau(T,\nu^{-1/2}\sigma)$	$L(s,\nu^{1/2}\sigma)^2$
	c	$L(\nu^{1/2}St_{GL(2)},\nu^{-1/2}\sigma)$	$L(s,\nu^{1/2}\sigma)^2 L(s,\nu^{-1/2}\sigma)$
	d	$L(\nu,1_{F^\times} \rtimes \nu^{-1/2}\sigma)$	$L(s,\nu^{1/2}\sigma)^2 L(s,\nu^{-1/2}\sigma)^2$
VII		$\chi \rtimes \pi$	1
VIII	a	$\tau(S,\pi)$	1
	b	$\tau(T,\pi)$	1
IX	a	$\delta(\nu\xi,\nu^{-1/2}\pi)$	1
	b	$L(\nu\xi,\nu^{-1/2}\pi)$	1
X		$\pi \rtimes \sigma$	$L(s,\sigma)L(s,\omega_\pi\sigma)$
XI	a	$\delta(\nu^{1/2}\pi,\nu^{-1/2}\sigma)$	$L(s,\nu^{1/2}\sigma)$
	b	$L(\nu^{1/2}\pi,\nu^{-1/2}\sigma)$	$L(s,\nu^{1/2}\sigma)L(s,\nu^{-1/2}\sigma)$

Exercise 7.11 *Use Table 7.4 of Hecke eigenvalues for the generic representations, and evaluate the L-function with the formula given in the theorem for as many representations as possible. Check that you do indeed get the L-factor given in Table 7.5. See Sect. A.9 of [81] for the values of (A), (B), and (C) in the last column of Table 7.4.*

Chapter 8
Bessel Models and Applications

Holomorphic Siegel modular forms F correspond to cuspidal automorphic representations π_F of $\mathrm{GSp}_4(\mathbb{A})$ that are not globally generic, i.e., they do not have a global Whittaker model. Hence, a substantial amount of literature and results regarding properties of, say, cuspidal automorphic representations of $\mathrm{GL}_n(\mathbb{A})$, are just not applicable to π_F. On the other hand, it is known that every irreducible cuspidal automorphic representation of $\mathrm{GSp}_4(\mathbb{A})$ always has some *Bessel* model. In this chapter, we will study the Bessel models for local and global representations of GSp_4. We will also consider an important application of Bessel models toward obtaining an integral representation of the degree 8 L-function of π_F twisted by an irreducible cuspidal representation of $\mathrm{GL}_2(\mathbb{A})$.

8.1 The L-Function $L(s, \pi_F \times \tau)$

To motivate Bessel models let us consider the following problem. Let $F \in S_k(\Gamma_2)$ be a Hecke eigenform, and let π_F be the corresponding irreducible cuspidal automorphic representation of $G(\mathbb{A})$. Recall that $G = \mathrm{GSp}_4$. Let τ be any irreducible cuspidal automorphic representation of $\mathrm{GL}_2(\mathbb{A})$. We want to study the degree 8 L-function $L(s, \pi_F \times \tau)$. This is defined as an Euler product over all primes. At a prime p, where τ_p is unramified, it is straightforward to write down the local L-factor. Suppose $\pi_p = \chi_1 \times \chi_2 \rtimes \sigma$ where χ_1, χ_2 and σ are unramified characters of $\mathbb{Q}_p^\times$. Let $\tau_p = \mu_1 \times \mu_2$ with unramified characters μ_1, μ_2. We have (see Sect. 3.7 of [30])

$$L_p(s, \pi_p \times \tau_p) = \prod_{i=1,2} L(s, \sigma\mu_i)L(s, \sigma\chi_1\mu_i)L(s, \sigma\chi_2\mu_i)L(s, \sigma\chi_1\chi_2\mu_i)$$

There are two aspects of interest of $L(s, \pi_F \times \tau)$ (in fact, for any L-function)—analytic properties and arithmetic properties.

© Springer Nature Switzerland AG 2019
A. Pitale, *Siegel Modular Forms*, Lecture Notes in Mathematics 2240,
https://doi.org/10.1007/978-3-030-15675-6_8

(i) *Analytic properties*: The L-function $L(s, \pi_F \times \tau)$ is defined as an Euler product that converges in some right half-plane. We are interested in knowing

- if the L-function has a meromorphic continuation to $\mathbb{C}$,
- the nature of the poles if they exist, and
- a functional equation relating the values of the L-function at s and $1 - s$.

Answers to these questions form the key ingredients of the *converse theorem* approach to the problem of Langlands transfer of π_F to $GL_4(\mathbb{A})$. Langlands conjectures predict such a transfer exists, and, once established, one can use known properties of GL_4 automorphic representations to obtain new results for π_F.

Out of the local components π_p of π_F, it is quite straightforward to find appropriate local representations Π_p of $GL_4(\mathbb{Q}_p)$. One can use these to form a representation $\Pi = \otimes \Pi_p$ of $GL_4(\mathbb{A})$. We want to know if Π is a **cuspidal**, **automorphic** representation of $GL_4(\mathbb{A})$. A converse theorem due to Cogdell and Piatetski-Shapiro [20] states that indeed Π is a cuspidal automorphic representation if every element in the family of L-functions $\{L(s, \Pi \times \tau) : \tau$ automorphic representation of $GL_2(\mathbb{A})\}$ is *nice*, i.e., has analytic continuation to all of $\mathbb{C}$, is bounded on vertical strips, and satisfies a functional equation with respect to $s \mapsto 1 - s$.

The way Π is constructed out of π_F leads to the conclusion that $L(s, \Pi \times \tau) = L(s, \pi_F \times \tau)$ for all τ. When τ is not cuspidal, the *nice* properties of $L(s, \pi_F \times \tau)$ are obtained by Krieg and Raum [54]. In the non-cuspidal case, the L-function $L(s, \pi_F \times \tau)$ is essentially the product of two copies of the spin L-function of F twisted by characters.

Hence, we are reduced to studying the analytic properties of $L(s, \pi_F \times \tau)$ for cuspidal representations τ of $GL_2(\mathbb{A})$.

(ii) *Arithmetic properties*: Let τ correspond to a holomorphic modular form. Deligne's conjecture [21] predicts that $L(s, \pi_F \times \tau)$ evaluated at certain *critical points* (a finite set of integers or half-integers depending on the weights of the modular forms) are algebraic numbers up to certain prescribed transcendental periods. An example of such a special value result is the fact that

$$\frac{\zeta(2k)}{\pi^{2k}} \in \mathbb{Q}, \text{ for all } k \in \mathbb{N}.$$

Shimura proved the following special values theorem for elliptic modular forms.

Theorem 8.1 (Shimura [95]) *Let f be a primitive holomorphic cusp form of weight $k \geq 2$ and level N. Then there exist nonzero complex numbers $u(\varepsilon, f)$ for $\varepsilon \in \{0, 1\}$ such that, for any Hecke character χ, and any integer m satisfying $0 < m < k$, we have*

$$\frac{L_f(m, f, \chi)}{(2\pi i)^m \tau(\chi) u(\varepsilon, f)} \in \bar{\mathbb{Q}},$$

where ε is given by $\chi(-1)(-1)^m = (-1)^\varepsilon$; $\tau(\chi)$ is the Gauss sum attached to χ, and $L_f(s, f, \chi)$ is the finite part of the L-function of f twisted by χ.

There has been tremendous progress in recent years toward proving arithmeticity of special values of L-functions in various settings. At the same time, this is a very active area of current research with wide open problems yet to be solved. We would like to discuss this problem in details in the setting of special values of $L(s, \pi_F \times \tau)$.

One approach to studying $L(s, \pi_F \times \tau)$ is to obtain an integral representation for it. This was done by Furusawa in [30]. The essential idea is to construct an Eisenstein series $E(g, s; f)$ on a bigger group $GU(2, 2) \supset GSp_4$ using the representation τ. This is a function on $GU(2, 2)(\mathbb{A})$ depending on a section f in an induced representation obtained from τ and s, a complex number. Let ϕ be any cusp form in π_F. Define the integral

$$Z(s, f, \phi) := \int_{G(\mathbb{Q})Z_G(\mathbb{A})\backslash G(\mathbb{A})} E(g, s; f)\phi(g)dg. \tag{8.1}$$

Theorem 8.2 (Basic Identity) *We have*

$$Z(s, f, \phi) = \int_{R(\mathbb{A})\backslash G(\mathbb{A})} W_f(\eta g, s)B_\phi(g)dg,$$

where

$$W_f(g, s) = \int_{\mathbb{Q}\backslash\mathbb{A}} f(\begin{bmatrix} 1 & & & \\ & 1 & x & \\ & & 1 & \\ & & & 1 \end{bmatrix} g)\psi(cx)dx$$

and

$$B_\phi(g) = \int_{Z_G(\mathbb{A})R(\mathbb{Q})\backslash R(\mathbb{A})} \phi(rg)(\Lambda \otimes \theta)(r)^{-1}dr.$$

This is Theorem 2.4 of [30]. B_ϕ is the element in the global Bessel model for π_F corresponding to ϕ. We will discuss Bessel models next. As mentioned above, Siegel modular forms F of degree 2 correspond to cuspidal automorphic representations π_F of $GSp_4(\mathbb{A})$ that are **not** globally generic. This means that π_F does not have a global Whittaker model. Recall that a Whittaker model provides a model for a representation in terms of $\mathbb{C}$-valued functions which transform according to a character when translated by the unipotent radical of the Borel subgroup. An alternate model is the Bessel model. In principal, it is quite similar to the Whittaker model–Bessel model provides a model for the representation in terms of $\mathbb{C}$-valued functions which transform according to a character of the *Bessel* subgroup. Let us describe this now.

8.2 Definition of Global Bessel Model

Let $S = \begin{bmatrix} a & b/2 \\ b/2 & c \end{bmatrix} \in M_2(\mathbb{Q})$ and let the discriminant of S be defined by $d = d(S) = -\det(2S) = b^2 - 4ac$. Assume that S is anisotropic, hence d is not a square in $\mathbb{Q}$.

Exercise 8.3 *Let S be as above.*

(i) *Define $\xi = \xi_S = \begin{bmatrix} b/2 & c \\ -a & -b/2 \end{bmatrix}$. Show that $F(\xi) := \{x + y\xi : x, y \in \mathbb{Q}\}$ is a quadratic extension of $\mathbb{Q}$ in $M_2(\mathbb{Q})$.*

(ii) *Let $L = \mathbb{Q}(\sqrt{d})$ be the quadratic subfield of $\mathbb{C}$. Show that*

$$F(\xi) \ni x + y\xi \mapsto x + y\frac{\sqrt{d}}{2} \in L, \qquad x, y \in \mathbb{Q}$$

is an isomorphism.

(iii) *Let $T = T_S$ be the subgroup of GL_2 defined by*

$$T(\mathbb{Q}) := \{g \in \mathrm{GL}_2(\mathbb{Q}) : {}^t g S g = \det(g) S\}. \tag{8.2}$$

Show that

$$T(\mathbb{Q}) \simeq F(\xi)^\times,$$

Hence, $T(\mathbb{Q}) \simeq L^\times$.

We can embed T in GSp_4 as follows.

$$T \ni g \mapsto \begin{bmatrix} g & \\ & \det(g)\,{}^t g^{-1} \end{bmatrix} \in \mathrm{GSp}_4.$$

Let U denote the unipotent radical of the Siegel parabolic subgroup, i.e.,

$$U = \{u(X) = \begin{bmatrix} 1_2 & X \\ & 1_2 \end{bmatrix} : X = {}^t X\}.$$

Finally, we define the *Bessel subgroup* of GSp_4 by $R = TU$.

Let ψ be a nontrivial additive character of $\mathbb{Q}\backslash\mathbb{A}$. Define a character $\theta = \theta_S$ of $U(\mathbb{A})$ by $\theta(u(X)) = \psi(\mathrm{Tr}(SX))$. Let Λ be any character of $L^\times\backslash\mathbb{A}_L^\times$ thought of as a character of $T(\mathbb{Q})\backslash T(\mathbb{A})$. Then, we get a character $\Lambda \otimes \theta$ of $R(\mathbb{A})$ defined by

$$(\Lambda \otimes \theta)(tu) := \Lambda(t)\theta(u), \text{ for } t \in T(\mathbb{A}), u \in U(\mathbb{A}).$$

Exercise 8.4 *For $t \in T(\mathbb{A})$, $u \in U(\mathbb{A})$, show that*

$$(\Lambda \otimes \theta)(tu) = (\Lambda \otimes \theta)(ut).$$

Let π be an irreducible cuspidal automorphic representation of $GSp_4(\mathbb{A})$ and assume that $\Lambda|_{\mathbb{A}^\times} = \omega_\pi$, the central character of π. For any $\phi \in \pi$, define the function B_ϕ on $GSp_4(\mathbb{A})$ by

$$B_\phi(h) = \int\limits_{Z(\mathbb{A})R(\mathbb{Q})\backslash R(\mathbb{A})} (\Lambda \otimes \theta)(r)^{-1}\phi(rh)dr. \qquad (8.3)$$

Exercise 8.5 *Show that the B_ϕ satisfy the transformation property*

$$B_\phi(rh) = (\Lambda \otimes \theta)(r)B_\phi(h)$$

for all $r \in R(\mathbb{A})$ and $h \in GSp_4(\mathbb{A})$.

The $\mathbb{C}$-vector space spanned by $\{B_\phi : \phi \in \pi\}$ is called a *global Bessel model* of type (S, Λ, ψ) for π. It is known by the work of Li [58] that every π either has a Whittaker model or a Bessel model for some (S, Λ, ψ).

Let $F \in S_k(\Gamma_2)$ be a Hecke eigenform with Fourier coefficients $\{A(T) : T > 0\}$. Let π_F be the corresponding cuspidal automorphic representation of $GSp_4(\mathbb{A})$. In order to compute the zeta integral $Z(s, f, \phi)$ defined in (8.1) we are going to need a (S, Λ, ψ)-Bessel model with S and Λ satisfying certain special properties. We call the triple (S, Λ, ψ) of *fundamental type* if S and Λ satisfy the following two properties.

(i) $S = \begin{bmatrix} a & b/2 \\ b/2 & 1 \end{bmatrix}$ with $a, b \in \mathbb{Z}$ and $d = b^2 - 4a < 0$ is a fundamental discriminant.

(ii) The Hecke character Λ is unramified at all the finite places of $L = \mathbb{Q}(\sqrt{d})$ and is trivial at infinity, i.e., it is a character of the ideal class group of L.

Recall from (4.2)

$$R(F, L, \Lambda) = \sum_{c \in \mathrm{Cl}_L} A(c)\Lambda(c)^{-1}.$$

Saha [88] has proved that π_F has a global Bessel model of fundamental type (S, Λ, ψ) if and only if $R(F, L, \Lambda) \neq 0$. Hence, we have the following important result.

Theorem 8.6 (Saha [88, Remark 1.1]) *Let $F \in S_k(\Gamma_2)$ be a Hecke eigenform, and let π_F be the corresponding cuspidal automorphic representation of $GSp_4(\mathbb{A})$. Then, π_F has a global Bessel model of fundamental type.*

8.3   Local Bessel Model

Let $p < \infty$ be a finite prime. Consider the exact situation as above excepting that
we take all the elements in $\mathbb{Q}_p$ instead of $\mathbb{Q}$. We have two possibilities. Either d
is not a square in $\mathbb{Q}_p^\times$, in which case $L_p = \mathbb{Q}_p(\sqrt{d})$. Or, d is a square in $\mathbb{Q}_p^\times$, and
then $L_p = \mathbb{Q}_p \oplus \mathbb{Q}_p$. In the latter case, the map from $\mathbb{Q}_p(\xi)$ to L_p is given by
$x + y\xi \mapsto (x + y\sqrt{d}/2, x - y\sqrt{d}/2)$.

We define the Legendre symbol as

$$\left(\frac{L_p}{p}\right) = \begin{cases} -1 & \text{if } L_p/\mathbb{Q}_p \text{ is an unramified field extension,} \\ 0 & \text{if } L_p/\mathbb{Q}_p \text{ is a ramified field extension,} \\ 1 & \text{if } L_p = \mathbb{Q}_p \oplus \mathbb{Q}_p. \end{cases} \qquad (8.4)$$

These three cases are referred to as the *inert case*, *ramified case*, and *split case*,
respectively. If L_p is a field, then let $\mathfrak{o}_{L_p}$ be its ring of integers and $\mathfrak{p}_{L_p}$ be the maximal
ideal of $\mathfrak{o}_{L_p}$. If $L_p = \mathbb{Q}_p \oplus \mathbb{Q}_p$, then let $\mathfrak{o}_{L_p} = \mathbb{Z}_p \oplus \mathbb{Z}_p$. Let ϖ_{L_p} be a uniformizer
in $\mathfrak{o}_{L_p}$ if L_p is a field, and set $\varpi_{L_p} = (p, 1)$ if L_p is not a field. In the field case let
v_{L_p} be the normalized valuation on L_p.

Let $T(\mathbb{Q}_p)$ be the torus of $\mathrm{GL}_2(\mathbb{Q}_p)$ defined as in (8.2) and let $U(\mathbb{Q}_p)$ be the
unipotent radical of the Siegel parabolic subgroup of $\mathrm{GSp}_4(\mathbb{Q}_p)$. Let $R(\mathbb{Q}_p) =
T(\mathbb{Q}_p)U(\mathbb{Q}_p)$ be the Bessel subgroup of $\mathrm{GSp}_4(\mathbb{Q}_p)$. Let ψ_p have conductor $\mathbb{Z}_p$,
and let θ_p be the character of $U(\mathbb{Q}_p)$ given by $\theta_p(u(X)) = \psi_p(\mathrm{Tr}(SX))$. Let Λ_p be a
character of $T(\mathbb{Q}_p) \simeq L_p^\times$. We have a character, denoted by $\Lambda_p \otimes \theta_p$, on $R(\mathbb{Q}_p)$ given
by $tu \mapsto \Lambda_p(t)\theta_p(u)$. Let $\mathcal{S}(\Lambda_p, \theta_p)$ be the space of all locally constant functions
$B : \mathrm{GSp}_4(\mathbb{Q}_p) \to \mathbb{C}$ with the *Bessel transformation property*

$$B(rg) = (\Lambda_p \otimes \theta_p)(r)B(g) \qquad \text{for all } r \in R(\mathbb{Q}_p) \text{ and } g \in \mathrm{GSp}_4(\mathbb{Q}_p). \qquad (8.5)$$

If an irreducible, admissible representation (π_p, V) of $\mathrm{GSp}_4(\mathbb{Q}_p)$ is isomorphic
to a subrepresentation of $\mathcal{S}(\Lambda_p, \theta_p)$, then this realization of π_p is called a (Λ_p, θ_p)-
Bessel model.

Exercise 8.7 *Let (π_p, V) be an irreducible admissible representation of* $\mathrm{GSp}_4(\mathbb{Q}_p)$.
A linear functional ℓ on V is called a (Λ_p, θ_p)-Bessel functional if it satisfies

$$\ell(\pi_p(tu)v) = \Lambda_p(t)\theta_p(u)\ell(v), \qquad \text{for all } t \in T(\mathbb{Q}_p), u \in U(\mathbb{Q}_p), v \in V.$$

(i) Given a (Λ_p, θ_p)-Bessel functional ℓ, show that the map

$$V \ni v \mapsto B_v, \qquad B_v(g) := \ell(\pi(g)v)$$

gives a (Λ_p, θ_p)-Bessel model for π_p.

(ii) Suppose π_p has a (Λ_p, θ_p)-Bessel model then show that the linear functional defined by

$$\ell(B) := B(1)$$

is a (Λ_p, θ_p)-Bessel functional.

This exercise shows that π_p has a (Λ_p, θ_p)-Bessel model if and only if it has a (Λ_p, θ_p)-Bessel functional.

Prasad and Takloo-Bighash [78] figured out which non-supercuspidal representations of $GSp_4(\mathbb{Q}_p)$ have Bessel models for which Λ_p. This was reproved and some gaps were filled in by Roberts and Schmidt [82]. In Table 8.1 we have the information compiled for all the representations that are induced from the Borel subgroup.

In addition to the criteria of existence, another very important result is that the Bessel model, when it exists, is unique.

For the archimedean case, suppose π_∞ is the lowest weight representation of $GSp_4(\mathbb{R})$ corresponding to a scalar valued weight k Siegel modular form. There are two options for L_∞—either $L_\infty = \mathbb{C}$ (the non-split case) or $L_\infty = \mathbb{R} \oplus \mathbb{R}$

Table 8.1 The Bessel models of the irreducible, admissible representations of $GSp_4(\mathbb{Q}_p)$ that can be obtained via induction from the Borel subgroup. The symbol N stands for the norm map from $L^\times \cong T(\mathbb{Q}_p)$ to $\mathbb{Q}_p^\times$

		Representation	(Λ, θ)-Bessel functional exists exactly for ...	
			$L = \mathbb{Q}_p \oplus \mathbb{Q}_p$	$L/\mathbb{Q}_p$ a field extension
I		$\chi_1 \times \chi_2 \rtimes \sigma$ (irreducible)	all Λ	all Λ
II	a	$\chi St_{GL(2)} \rtimes \sigma$	all Λ	$\Lambda \neq (\chi\sigma) \circ N$
	b	$\chi 1_{GL(2)} \rtimes \sigma$	$\Lambda = (\chi\sigma) \circ N$	$\Lambda = (\chi\sigma) \circ N$
III	a	$\chi \rtimes \sigma St_{GSp(2)}$	all Λ	all Λ
	b	$\chi \rtimes \sigma 1_{GSp(2)}$	$\Lambda(\text{diag}(a, b, b, a))=$ $\chi(a)\sigma(ab)$ or $\chi(b)\sigma(ab)$	—
IV	a	$\sigma St_{GSp(4)}$	all Λ	$\Lambda \neq \sigma \circ N$
	b	$L(\nu^2, \nu^{-1}\sigma St_{GSp(2)})$	$\Lambda = \sigma \circ N$	$\Lambda = \sigma \circ N$
	c	$L(\nu^{3/2}St_{GL(2)}, \nu^{-3/2}\sigma)$	$\Lambda(\text{diag}(a, b, b, a))=$ $\nu(ab^{-1})\sigma(ab)$ or $\nu(a^{-1}b)\sigma(ab)$	—
	d	$\sigma 1_{GSp(4)}$	—	—
V	a	$\delta([\xi, \nu\xi], \nu^{-1/2}\sigma)$	all Λ	$\Lambda \neq \sigma \circ N$, $\Lambda \neq (\xi\sigma) \circ N$
	b	$L(\nu^{1/2}\xi St_{GL(2)}, \nu^{-1/2}\sigma)$	$\Lambda = \sigma \circ N$	$\Lambda = \sigma \circ N$, $\Lambda \neq (\xi\sigma) \circ N$
	c	$L(\nu^{1/2}\xi St_{GL(2)}, \xi\nu^{-1/2}\sigma)$	$\Lambda = (\xi\sigma) \circ N$	$\Lambda \neq \sigma \circ N$, $\Lambda = (\xi\sigma) \circ N$
	d	$L(\nu\xi, \xi \rtimes \nu^{-1/2}\sigma)$	—	$\Lambda = \sigma \circ N$, $\Lambda = (\xi\sigma) \circ N$
VI	a	$\tau(S, \nu^{-1/2}\sigma)$	all Λ	$\Lambda \neq \sigma \circ N$
	b	$\tau(T, \nu^{-1/2}\sigma)$	—	$\Lambda = \sigma \circ N$
	c	$L(\nu^{1/2}St_{GL(2)}, \nu^{-1/2}\sigma)$	$\Lambda = \sigma \circ N$	—
	d	$L(\nu, 1_{F^\times} \rtimes \nu^{-1/2}\sigma)$	$\Lambda = \sigma \circ N$	—

(the split case). In [74] Pitale and Schmidt worked out the criteria for existence and uniqueness of Bessel models in this situation. We have the following answer.

 (i) In the split case, π_∞ *never* has a Bessel model for any character Λ_∞.
 (ii) In the non-split case, any character Λ_∞ of $\mathbb{C}^\times$ is given by the pair $(s, m) \in \mathbb{C} \times \mathbb{Z}$ by the formula

$$\Lambda_\infty(\gamma e^{2\pi i \delta}) = \gamma^s e^{2\pi i m \delta}, \text{ where } \gamma, \delta \in \mathbb{R}, \gamma > 0.$$

Then π_∞ has a $(\Lambda_\infty, \theta_\infty)$-Bessel model if and only if $m = 0$. When a Bessel model exists, it is unique.

8.4 Explicit Formulas for Distinguished Vectors in Local Bessel Models

Let us for the moment go back to the global Bessel model. Given $\phi \in \pi_F$, we have the global Bessel function B_ϕ given by (8.3).

Exercise 8.8 *Show that, if we evaluate B_ϕ at $h = (1, \cdots, h_p, \cdots, 1, \cdots)$, then it is a function on $\mathrm{GSp}_4(\mathbb{Q}_p)$ satisfying (8.5).*

The above exercise states that if a global (S, Λ, ψ)-Bessel model exists for π, then a local (S, Λ_p, ψ_p)-Bessel model exists for π_p. In fact, there is a canonical isomorphism between the global (S, Λ, ψ)-Bessel model and the restricted tensor product of the local (S, Λ_p, ψ_p)-Bessel models. If $(B_p)_p$ is a collection of local Bessel functions B_p such that $B_p|_{K_p} = 1$ for almost all p, then this isomorphism is such that $\otimes_p B_p$ corresponds to the global function

$$B((h_p)_p) = \prod_{p \leq \infty} B_p(h_p).$$

Let $\pi = \pi_F$ be the representation corresponding to a Hecke eigenform $F \in S_k(\Gamma_2)$. Then, for every $p < \infty$, the representation π_p is a unramified.

Exercise 8.9 *Take $\phi = \Phi_F$. Let $B_{\Phi_F} = \prod_p B_p$. Then show that, for $p < \infty$, the vector B_p is the spherical vector in π_p, i.e., B_p is right invariant under K_p.*

Recall that we had the global integral

$$Z(s, f, \phi) = \int_{R(\mathbb{A})\backslash G(\mathbb{A})} W_f(\eta h, s) B_\phi(h) dh.$$

Now we can conclude that $Z(s, f, \phi)$ is Eulerian.

$$Z(s, f, \phi) = \prod_{p \leq \infty} Z_p(s),$$

where

$$Z_p(s) := \int_{R(\mathbb{Q}_p) \backslash G(\mathbb{Q}_p)} W_p(\eta h, s) B_p(h) dh.$$

Here, we have also used the uniqueness of Whittaker models for GL$_2$ to write W_f as a product of local functions W_p. Also, in the archimedean case, we replace $\mathbb{Q}_p$ by $\mathbb{R}$ above.

Since we got a Euler product for the integral, we now hope that each of the local integrals computes to a local factor of a L-function. The only way to find that out is to do an explicit computation. To do these computations, we need explicit formulas for the W_p and B_p. We can obtain the formulas for W_p from the formulas for vectors in the Whittaker models for GL$_2$. Let us discuss the case for B_p.

Let B_p be the spherical vector in the (Λ_p, θ_p)-Bessel model of π_p. We have the following double coset decomposition for GSp$_4(\mathbb{Q}_p)$ (see Eq. 3.4.2 of [30]).

$$\mathrm{GSp}_4(\mathbb{Q}_p) = \bigsqcup_{l \in \mathbb{Z}, m \geq 0} R(\mathbb{Q}_p) h_p(l, m) K_p, \quad h_p(l, m) := \mathrm{diag}(p^{l+2m}, p^{l+m}, 1, p^m).$$

Exercise 8.10 *Let B_p be a spherical vector in the (Λ_p, θ_p)-Bessel model of π_p.*

(i) *Show that B_p is completely determined by its values on $h_p(l, m)$ for $l \in \mathbb{Z}, m \geq 0$.*

(ii) *Use the fact that $a, b \in \mathbb{Z}_p, c \in \mathbb{Z}_p^\times$, and that ψ_p has conductor $\mathbb{Z}_p$ to show that $B_p(h_p(l, m)) = 0$ if $l < 0$.*

Sugano [101] obtained explicit formulas for B_p. Let us state them here. Let π_p be the spherical constituent of $\chi_1 \times \chi_2 \rtimes \sigma$. Set

$$\gamma_1 = \chi_1(p)\chi_2(p)\sigma(p), \quad \gamma_2 = \chi_1(p)\sigma(p), \quad \gamma_3 = \sigma(p), \quad \gamma_4 = \chi_2(p)\sigma(p).$$

Set

$$\epsilon_p = \begin{cases} 0 & \text{if } \left(\frac{L_p}{p}\right) = -1; \\ \Lambda_p(\varpi_{L_p}) & \text{if } \left(\frac{L_p}{p}\right) = 0; \\ \Lambda_p(\varpi_{L_p}) + \Lambda_p(p\varpi_{L_p}^{-1}) & \text{if } \left(\frac{L_p}{p}\right) = 1. \end{cases}$$

Define the generating function

$$C_p(x, y) := \sum_{l \geq 0} \sum_{m \geq 0} B_p(h_p(l, m)) x^m y^l.$$

Theorem 8.11 (Sugano [101]) *With notations as above, we have*

$$C_p(x, y) = \frac{H_p(x, y)}{P_p(x)Q_p(y)}.$$

Here

$$P_p(x) = (1 - \gamma_1\gamma_2 p^{-2}x)(1 - \gamma_1\gamma_4 p^{-2}x)(1 - \gamma_2\gamma_3 p^{-2}x)(1 - \gamma_3\gamma_4 p^{-2}x)$$

$$Q_p(y) = \prod_{i=1}^{4}(1 - \gamma_i p^{-3/2}y)$$

$$H_p(x, y) = (1 - A_2 A_3 xy^2)\big(M_1(x)(1 + A_2 x) + A_2 A_5 A_1^{-1}\alpha x^2\big)$$

$$\qquad - A_2 xy(\alpha M_1(x) - A_5 M_2(x)) - A_5 P_p(x)y - A_2 A_4 P_p(x)y^2$$

$$M_1(x) = 1 - A_1^{-1}(A_1 + A_4)^{-1}(A_1 A_5\alpha + A_4\beta - A_1 A_5^2 - 2A_1 A_2 A_4)x$$

$$\qquad + A_1^{-1}A_2^2 A_4 x^2$$

$$M_2(x) = 1 + A_1^{-1}(A_1 A_2 - \beta)x + A_1^{-1}A_2(A_1 A_2 - \beta)x^2 + A_2^3 x^3$$

$$\alpha = p^{-3/2}\sum_{i=1}^{4}\gamma_i, \quad \beta = p^{-3}\sum_{1\le i<j\le 4}\gamma_i\gamma_j, \quad A_1 = p^{-1}, \quad A_2 = p^{-2}\Lambda_p(p)$$

$$A_3 = p^{-3}\Lambda_p(p), \quad A_4 = -p^{-2}\Big(\frac{L_p}{p}\Big), \quad A_5 = p^{-2}\epsilon_p$$

Exercise 8.12 *Assume that $\Lambda_p = 1$, and that one of the γ_i above is $p^{\pm 1/2}$. Then show that, for all $l, m \ge 0$, we have*

$$B_p(h_p(l, m)) = \sum_{i=0}^{l} p^{-i} B_p(h_p(0, l + m - i)). \tag{8.6}$$

We know that, if F is a Saito–Kurokawa lift then its Fourier coefficients satisfy the recurrence relation (2.2). We also know that, in this case, the local representation π_p is of type IIb, which satisfies the hypothesis that one of the γ_i is $p^{\pm 1/2}$. The recursion formula (8.6) for the B_p is a local analogue of the Maass relations (2.2). In [70, Theorem 7.1], Pitale–Saha–Schmidt showed that these local recurrence relations (8.6) do imply the global Maass relations (2.2), thereby finally giving a representation theoretic explanation for the existence of the Maass relations (2.2).

Exercise 8.13 *Show that*

$$\frac{C_p(0, p^{-s})}{H_p(0, p^{-s})} = L(s + 3/2, \pi_p, \text{spin}), \qquad \frac{C_p(p^{-s}, 0)}{(1 - p^{-s-2})H_p(p^{-s}, 0)} = L(s + 2, \pi_p, \text{std}),$$

$$\frac{C_p(p^{-s}, p^{-s})}{(1 - p^{-s-2})H_p(p^{-s}, p^{-s})} = L(s + 3/2, \pi_p, \text{spin})L(s + 2, \pi_p, \text{std}).$$

With the values of B_p in hand, we can compute the local integral Z_p.

Theorem 8.14 [69, Theorem 2.2.1] *The local zeta integral $Z_p(s)$ is given by*

$$Z_p(s) = \frac{L(3s + \frac{1}{2}, \tilde{\pi}_p \times \tilde{\tau}_p)}{L(6s + 1, \Lambda_0|_{\mathbb{Q}_p^\times})L(3s + 1, \tau_p \times \text{AI}(\Lambda_p) \times \Lambda_0|_{\mathbb{Q}_p^\times})} Y(s), \qquad (8.7)$$

where

$$Y(s) = \begin{cases} 1 & \text{if } \tau_p = \beta_1 \times \beta_2, \ \beta_1, \beta_2 \text{ unramified,} \\ L(6s + 1, \Lambda_0|_{\mathbb{Q}_p^\times}) & \text{if } \tau_p = \beta_1 \times \beta_2, \ \beta_1 \text{ unram., } \beta_2 \text{ ram.,} \\ & \left(\frac{L}{\mathfrak{p}}\right) = \pm 1, \text{ OR } \tau = \beta_1 \times \beta_2, \ \beta_1 \text{ unram.,} \\ & \beta_2 \text{ ram., } \left(\frac{L}{\mathfrak{p}}\right) = 0 \text{ and } \beta_2 \chi_{L_p/\mathbb{Q}_p} \text{ ramified,} \\ & \text{OR } \tau = \Omega \text{St}_{\text{GL}(2)}, \ \Omega \text{ unramified,} \\ \dfrac{L(6s + 1, \Lambda_0|_{\mathbb{Q}_p^\times})}{1 - \Lambda_p(\varpi_{L_p})(\omega_{\pi_p}\beta_2)^{-1}(p)p^{-3s-1}} & \text{if } \tau_p = \beta_1 \times \beta_2, \ \beta_1 \text{ unram., } \beta_2 \text{ ram.,} \\ & \left(\frac{L_p}{\mathfrak{p}}\right) = 0, \text{ and } \beta_2 \chi_{L_p/\mathbb{Q}_p} \text{ unramified,} \\ L(6s + 1, \Lambda_0|_{\mathbb{Q}_p^\times}) & \text{if } \tau_p = \beta_1 \times \beta_2, \ \beta_1, \beta_2 \text{ ramified,} \\ \times L(3s + 1, \tau_p \times \text{AI}(\Lambda_p) \times \Lambda_0|_{\mathbb{Q}_p^\times}) & \text{OR } \tau_p = \Omega \text{St}_{\text{GL}(2)}, \ \Omega \text{ ramified,} \\ & \text{OR } \tau_p \text{ supercuspidal.} \end{cases}$$

In (8.7), $\tilde{\pi}_p$ and $\tilde{\tau}_p$ denote the contragredient of π_p and τ_p, respectively. The symbol $\text{AI}(\Lambda_p)$ stands for the $\text{GL}_2(\mathbb{Q}_p)$ representation attached to the character Λ_p of $L_p^\times$ via automorphic induction, and $\chi_{L_p/\mathbb{Q}_p}$ stands for the quadratic character of $\mathbb{Q}_p^\times$ associated with the extension $L_p/\mathbb{Q}_p$. Λ_0 is a character of $L^\times\backslash\mathbb{A}_L^\times$ associated to Λ and ω_τ. The function $L(3s + 1, \tau_p \times \text{AI}(\Lambda_p) \times \Lambda_0|_{\mathbb{Q}_p^\times})$ is a standard L-factor for $\text{GL}_2 \times \text{GL}_2 \times \text{GL}_1$.

In the archimedean case, the weight k vector in the Bessel model of π_∞ is given by the following formula. For $h_\infty \in \text{GSp}_4(\mathbb{R})^+$, we have (see Eq. 4.3.4 of [30])

$$B_\infty(h_\infty) = \mu(h_\infty)^k \overline{\det(J(h_\infty, I))}^{-k} e^{-2\pi i \text{Tr}(S\, h_\infty\langle I \rangle)} R(F, L, \Lambda).$$

We get the global integral representation theorem.

Theorem 8.15 [69, Theorem 2.3.2] *We have*

$$Z(s, f, \phi) = \frac{L(3s + \frac{1}{2}, \tilde{\pi} \times \tilde{\tau})}{L(6s + 1, \Lambda_0|_{\mathbb{A}^\times}) L(3s + 1, \tau \times \mathrm{AI}(\Lambda) \times \Lambda_0|_{\mathbb{A}^\times})} \cdot B_\phi(1) \prod_{p \leq \infty} Y_p(s),$$

(8.8)

where, for $p < \infty$, the $Y_p(s)$ is given in Theorem 8.14, and we have

$$Y_\infty(s) = \pi \overline{R(F, L, \Lambda)} (4\pi)^{-3s - 3k/2 + 3/2} (\sqrt{d})^{-6s - k} \frac{\Gamma(3s + 3k/2 - 3/2)}{6s + k - 1}.$$

Chapter 9
Analytic and Arithmetic Properties of $GSp_4 \times GL_2$ L-Functions

In this chapter, we will start from the integral representation for $L(s, \pi_F \times \tau)$ obtained in the previous chapter. We will use this integral representation to obtain analytic and arithmetic properties of the L-functions. We will also present several applications.

9.1 Functional Equation and Analytic Continuation

Let us recall from the previous chapter that $F \in S_k(\Gamma_2)$ is a Hecke eigenform, and π_F is the irreducible cuspidal automorphic representation of $GSp_4(\mathbb{A})$ corresponding to F. Let τ be an irreducible cuspidal automorphic representation of $GL_2(\mathbb{A})$. In Theorem 8.15, we obtained an integral representation for $L(s, \pi_F \times \tau)$ in terms of the zeta integral $Z(s, f, \phi)$ defined in (8.1) using an Eisenstein series $E(g, s; f)$ on $GU(2, 2)(\mathbb{A})$ and a cusp form $\phi \in \pi_F$.

The Eisenstein series $E(g, s; f)$ has a functional equation and a meromorphic continuation to $\mathbb{C}$ (see Eq. 1.3.3 of [30]). Via the integral $Z(s, f, \phi)$, these properties can now be transferred to the L-function.

Theorem 9.1 [69, Theorem 2.4.3] *Let τ be such that τ_p is unramified for $p|d$. We have*

$$L(s, \pi_F \times \tau) = \epsilon(s, \pi_F \times \tau)L(1 - s, \tilde{\pi}_F \times \tilde{\tau}).$$

Recall that d is defined in the previous chapter with $L = \mathbb{Q}(\sqrt{d})$. To get the analytic continuation, we have to first get another integration formula which involves taking a degenerate Eisenstein series on $GU(3, 3)$, restricting it to the subgroup $GU(2, 2) \times GU(1, 1)$, and then integrating against a cusp form on GSp_4 and GL_2. This once again computes to $L(s, \pi_F \times \tau)$ up to certain known factors. The advantage of this new integral representation is that we have much more precise

© Springer Nature Switzerland AG 2019
A. Pitale, *Siegel Modular Forms*, Lecture Notes in Mathematics 2240,
https://doi.org/10.1007/978-3-030-15675-6_9

information regarding the nature and location of the poles of the degenerate Eisenstein series. Actually, getting to the conclusion that the L-function does not have any poles requires the application of the regularized Siegel–Weil formula due to Ichino [41]. This leads to the following theorem.

Theorem 9.2 [69, Theorem 4.1.1] *Let $F \in S_k(\Gamma_2)$ be a Hecke eigenform, which is not a Saito–Kurokawa lift. Let τ be an irreducible cuspidal automorphic representation of $GL_2(\mathbb{A})$ such that τ_p is unramified for $p|d$. Then $L(s, \pi_F \times \tau)$ is an entire function.*

9.2 Transfer to GL₄

We can write down explicitly the local parameters of π_p for all $p \leq \infty$. These are maps from the local Weil group to $GSp_4(\mathbb{C})$. Since π_F has a trivial central character, the image is in $Sp_4(\mathbb{C})$. The natural inclusion of $Sp_4(\mathbb{C})$ in $GL_4(\mathbb{C})$ gives us local parameters for GL_4. Since local Langlands correspondence is known for GL_4, we obtain local irreducible admissible representations Π_p of $GL_4(\mathbb{Q}_p)$ for every $p \leq \infty$. Define

$$\Pi_4 := \otimes_p \Pi_p.$$

This is an irreducible admissible representation of $GL_4(\mathbb{A})$ with the property that

$$L(s, \Pi_4 \times \tau) = L(s, \pi_F \times \tau)$$

for any irreducible cuspidal automorphic representation τ of $GL_i(\mathbb{A})$ for $i = 1, 2$.

Theorem 9.3 [69, Theorem 5.1.2] *Let π_F be a cuspidal automorphic representation of $GSp_4(\mathbb{A})$ as above, associated to a Siegel Hecke eigenform $F \in S_k(\Gamma_2)$. We assume that F is not of Saito–Kurokawa type. Then the admissible representation Π_4 of $GL_4(\mathbb{A})$ defined above is cuspidal automorphic. Hence Π_4 is a strong functorial lifting of π_F. This representation is symplectic, i.e., the exterior square L-function $L(s, \Pi_4, \Lambda^2)$ has a pole at $s = 1$.*

The exterior square L-function is discussed in details below. Since, in Theorems 9.1 and 9.2, we have some restriction on the representations τ that we are allowed to twist with, the converse theorem only gives us a weak functorial lift. This means that we obtain $\Pi' = \otimes_p \Pi'_p$, an automorphic representation of $GL_4(\mathbb{A})$, such that $\Pi_p \simeq \Pi'_p$ for all $p \nmid d$. To get the stronger fact that indeed Π_4 is automorphic, or that $\Pi'_p \simeq \Pi_p$ even for $p|d$, we use the fact that F is full level, and we use the rather strong theorem of Weissauer [106] of the validity of the generalized Ramanujan conjecture for non-Saito–Kurokawa lifts.

The argument for cuspidality is rather easy and will be left for the next exercise.

Exercise 9.4 *Suppose that* Π_4 *is not cuspidal. Then, it is known that* Π_4 *is a constituent of a globally induced representation from a proper parabolic subgroup of* GL$_4$.

(i) *Show that the inducing data for the globally induced representation mentioned above has to be unramified for all finite primes.*

(ii) *Show that* $L(s, \Pi_4)$ *will factor into product of global L-functions, and write down the 4 possible factorizations depending on the proper parabolic subgroups.*

(iii) *Note that* $\zeta(s)$ *has a pole, and* $L(s, \tau \times \tilde{\tau})$ *has a pole for any cuspidal representation of* GL$_2(\mathbb{A})$. *Use this in conjunction with Theorem 9.2, and the fact that* F *is not a Saito–Kurokawa lift, to get a contradiction.*

Hence, Π_4 *is cuspidal.*

Arthur [5] has proven Langlands functorial transfer from several classical groups to GL$_n$ using the trace formula. Since the representation π_F has a trivial central character, it is in fact a representation for PGSp$_4(\mathbb{A})$. We have an accidental isomorphism PGSp$_4 \simeq$ SO$_5$. Arthur's results can be applied to odd orthogonal groups and hence, the transfer from Theorem 9.3 can be deduced from the trace formula result.

9.3 Other Analytic Applications

We will list some more analytic results that can be deduced out of the transfer of π_F to GL$_4$.

Globally generic representation in the same L-packet: Since the representation Π_4 is symplectic, the work on backwards lifting by Ginzburg et al. [37] gives the existence of a globally generic cuspidal automorphic representation $\pi^g = \otimes_p \pi_p^g$ of GSp$_4(\mathbb{A})$ such that $\pi_p^g \simeq \pi_p$ for all finite primes p, and such that π_∞^g is the generic discrete series representation of PGSp$_4(\mathbb{R})$ lying in the same L-packet as π_∞. Any globally generic, cuspidal automorphic representation $\sigma \cong \otimes \sigma_p$ of GSp$_4(\mathbb{A})$ such that $\sigma_p \cong \pi_p$ for almost all p coincides with π^g.

Langlands transfer of π_F *to* GL$_5$: Let $\Pi \cong \otimes \Pi_p$ be an irreducible cuspidal representation of GL$_n(\mathbb{A})$ such that Π_p is unramified for every $p < \infty$. Let $\alpha_{1,p}, \cdots, \alpha_{n,p}$ be the Satake p-parameters of Π_p. The exterior square L-function $L(s, \Pi, \Lambda^2)$ and the symmetric square L-function $L(s, \Pi, \mathrm{Sym}^2)$ are defined by the following Euler products.

$$L(s, \Pi, \Lambda^2) = \prod_p \prod_{1 \le i < j \le n} (1 - \alpha_{i,p}\alpha_{j,p}p^{-s})^{-1}$$

$$L(s, \Pi, \mathrm{Sym}^2) = \prod_p \prod_{1 \le i \le j \le n} (1 - \alpha_{i,p}\alpha_{j,p}p^{-s})^{-1}$$

Exercise 9.5 *In this exercise, we will compute the symmetric square and exterior square L-functions of genus 1 and 2 Siegel modular forms.*

(1) *Let $f \in S_k(\Gamma_1)$ be a Hecke eigenform, and let $\tau_f \cong \otimes \tau_p$ be the irreducible cuspidal automorphic representation of $GL_2(\mathbb{A})$ corresponding to f. For every $p < \infty$, it is known that τ_p is unramified and the Satake p-parameters are given by $\alpha_p, \bar{\alpha}_p$ for some $\alpha_p \in \mathbb{C}$ with $|\alpha_p| = 1$. Show that $L(s, \tau_f, \Lambda^2) = \zeta(s)$. Also, find a formula for $L(s, \tau_f, Sym^2)$.*

(2) *Let Π_4 be the irreducible cuspidal automorphic representation of $GL_4(\mathbb{A})$ obtained from the Langlands transfer of $\pi_F \cong \otimes \pi_p$. The Satake p-parameters for π_p are given by $b_0, b_0b_1, b_0b_2, b_0b_1b_2$ as defined in Sect. 6.4. Using $b_0^2 b_1 b_2 = 1$, show that $L(s, \Pi_4, \Lambda^2) = L(s, \pi_F, std)\zeta(s)$.*

Kim [45] has shown that, given a representation Π of $GL_4(\mathbb{A})$, there is an automorphic representation of $GL_6(\mathbb{A})$, whose L-function is the exterior square L-function $L(s, \Pi, \Lambda^2)$. Also, such representations are obtained as $\mathrm{Ind}(\tau_1 \otimes \cdots \otimes \tau_k)$, where τ_i is a cuspidal automorphic representation of $GL_{n_i}(\mathbb{A})$. In our case, we have the relation $L(s, \Pi_4, \Lambda^2) = L(s, \pi_F, std)\zeta(s)$. Using the fact that $L(s, \pi_F, std)$ does not have a pole at $\mathrm{Re}(s) = 1$, we can show that there exists a cuspidal, automorphic representation Π_5 of $GL_5(\mathbb{A})$ such that

$$L(s, \pi_F, std) = L(s, \Pi_5)$$

(equality of completed Euler products). The representation Π_5 is a strong functorial lifting of π_F to GL_5 with respect to the morphism $\rho_5 : Sp_4(\mathbb{C}) \to GL_5(\mathbb{C})$ of dual groups. Moreover, Π_5 is orthogonal, i.e., the symmetric square L-function $L(s, \Pi_5, Sym^2)$ has a pole at $s = 1$.

Analytic properties of L-functions associated to π_F: Let ρ_n be the n-dimensional irreducible representation of $Sp_4(\mathbb{C})$ for values of n in the set $\{1, 4, 5, 10, 14, 16\}$. In the notation of Fulton and Harris [29, Sect. 16.2], we have $\rho_4 = \Gamma_{1,0}$, $\rho_5 = \Gamma_{0,1}$, $\rho_{10} = \Gamma_{2,0}$, $\rho_{14} = \Gamma_{0,2}$ and $\rho_{16} = \Gamma_{1,1}$. Note that ρ_4 is the spin representation, ρ_5 is the standard representation, and ρ_{10} is the adjoint representation of $Sp_4(\mathbb{C})$ on its Lie algebra. We have the following relations between these various representations – $\Lambda^2 \rho_4 = \rho_1 + \rho_5$, $\Lambda^2 \rho_5 = Sym^2 \rho_4 = \rho_{10}$, $Sym^2 \rho_5 = \rho_1 + \rho_{14}$, $\rho_4 \otimes \rho_5 = \rho_4 + \rho_{16}$. Since Π_4 and Π_5 are cuspidal automorphic representations of $GL_4(\mathbb{A})$ and $GL_5(\mathbb{A})$, we know a lot of analytic properties for the Sym^2, Λ^2 and the tensor product L-functions for Π_4 and Π_5. This allows us to conclude the following (see Theorem 5.2.1 of [69]).

The Euler products defining the finite part of the L-functions $L_f(s, \pi_F, \rho_n)$, for $n \in \{4, 5, 10, 14, 16\}$, are absolutely convergent for $\mathrm{Re}(s) > 1$. They have meromorphic continuation to the entire complex plane, have no zeros or poles on $\mathrm{Re}(s) \geq 1$, and the completed L-functions satisfy the functional equation

$$L(s, \pi_F, \rho_n) = \varepsilon(s, \pi_F, \rho_n)L(1 - s, \pi_F, \rho_n).$$

Furthermore, for $n \in \{4, 5, 10\}$, the functions $L(s, \pi_F, \rho_n)$ are entire and bounded in vertical strips.

Analytic properties of Rankin–Selberg L-functions: Let r be a positive integer, and τ a cuspidal, automorphic representation of $GL_r(\mathbb{A})$. Let σ_r be the standard representation of the dual group $GL_r(\mathbb{C})$. Then we can consider the Rankin–Selberg Euler products $L(s, \pi_F \times \tau, \rho_n \otimes \sigma_r)$, where ρ_n is one of the irreducible representations of $Sp_4(\mathbb{C})$ considered above. For $n = 4$ or $n = 5$, since Π_4 and Π_5 are functorial liftings of π_F, we have

$$L(s, \pi_F \times \tau, \rho_n \times \sigma_r) = L(s, \Pi_n \times \tau),$$

where the L-function on the right is a standard Rankin–Selberg L-function for $GL_n \times GL_r$. From the well-known properties of these L-functions, we have the following result (see Theorem 5.2.2 of [69]).

The Euler products defining the $GSp_4 \times GL_r$ L-functions $L(s, \pi_F \times \tau, \rho_n \otimes \sigma_r)$ are absolutely convergent for $\mathrm{Re}(s) > 1$. They have meromorphic continuation to the entire complex plane, and the completed L-functions satisfy the functional equation

$$L(s, \pi_F \times \tau, \rho_n \otimes \sigma_r) = \varepsilon(s, \pi_F \times \tau, \rho_n \otimes \sigma_r) L(1 - s, \tilde{\pi}_F \times \tilde{\tau}, \rho_n \otimes \sigma_r).$$

These L-functions are entire, bounded in vertical strips, and nonvanishing on $\mathrm{Re}(s) \geq 1$, except in the cases depending on the relation of τ with Π_4 or Π_5.

Analytic properties of $GSp_4 \times GSp_4$ L-functions: Let F and F' be Siegel cusp forms with respect to $Sp_4(\mathbb{Z})$. Assume that F and F' are Hecke eigenforms, that they are not Saito–Kurokawa lifts, and that π, resp. π', are the associated cuspidal, automorphic representations of $GSp_4(\mathbb{A})$. Let $n \in \{4, 5\}$ and $n' \in \{4, 5\}$. We have

$$L(s, \pi \times \pi', \rho_n \otimes \rho_{n'}) = L(s, \Pi_n \times \Pi'_{n'}).$$

Hence, the Euler products defining the $GSp_4 \times GSp_4$ L-functions $L(s, \pi \times \pi', \rho_n \otimes \rho_{n'})$ are absolutely convergent for $\mathrm{Re}(s) > 1$. They have meromorphic continuation to the entire complex plane, and the completed L-functions satisfy the expected functional equation. These functions are entire, bounded in vertical strips, and nonvanishing on $\mathrm{Re}(s) \geq 1$, except if $n = n'$ and $F = F'$ (see Theorem 5.2.3 of [69]).

Nonnegativity at $s = 1/2$ for L-functions: Lapid [56] has proved the nonnegativity of the central value $L(1/2, \pi \times \pi')$ for cuspidal automorphic representations π of GL_n and π' of $GL_{n'}$ satisfying certain hypothesis. Let F, F' be as above. Let χ be a Hecke character of $\mathbb{A}^\times$ (possibly trivial) such that $\chi^2 = 1$, τ_2 be a unitary, cuspidal, automorphic representation of $GL_2(\mathbb{A})$ with trivial central character, and τ_3 be a unitary, self-dual, cuspidal, automorphic representation of $GL_3(\mathbb{A})$. Then the central values

$$L(1/2, \pi \otimes \chi, \rho_4), \ L(1/2, \pi \otimes \tau_2, \rho_5 \otimes \sigma_2), \ L(1/2, \pi \otimes \tau_3, \rho_4 \otimes \sigma_3), \ L(1/2, \pi \times \pi', \rho_4 \otimes \rho_5),$$

are all nonnegative (see Theorem 5.2.4 of [69]).

9.4 Arithmetic Applications of the Integral Representation of $L(s, \pi_F \times \tau)$

The final application of the integral representation of $L(s, \pi_F \times \tau)$ concerns algebraicity results for its special values. Let us first discuss the critical points for the L-function.

Critical points: Suppose we have an L-function given by

$$L(s) = \prod_{p \leq \infty} L_p(s),$$

and suppose that $L(s)$ satisfies a functional equation with respect to $s \mapsto k - s$. Then, the *critical points* are the set of integers m for which both $L_\infty(s)$ and $L_\infty(k - s)$ **do not** have a pole at $s = m$. Let us consider the example of the Riemann zeta function. In this case, $L_\infty(s) = \pi^{-s/2}\Gamma(s/2)$. We know that the gamma function has a pole exactly at all nonpositive integers. Hence, we can conclude that, for an integer m, if $s = m$ is not a pole for both $L_\infty(s)$ and $L_\infty(1 - s)$, then m is either an even positive integer or an odd negative integer.

Exercise 9.6 *Suppose* $\Psi \in S_\ell(\Gamma_1)$ *has Fourier coefficients* $\{a(n)\}$. *Then, the completed L-function* $L(s, \Psi) := (2\pi)^{-s}\Gamma(s) \sum_{n>0} \frac{a(n)}{n^s}$ *satisfies the functional equation* $L(s, \Psi) = (-1)^{\ell/2}L(\ell - s, \Psi)$. *Show that the critical points in this case are all integers* m *satisfying* $0 < m < \ell$.

Assume that $\Psi \in S_k(\Gamma_0(N))$, with $k, N \in \mathbb{N}$, k even, is a Hecke eigenform. Let $\tau = \tau_\Psi$ be the cuspidal automorphic representation of $GL_2(\mathbb{A})$ corresponding to Ψ. Recall that we have a Hecke eigenform $F \in S_k(\Gamma_2)$. In this case

$$L_\infty(s, \pi_F \times \tau_\Psi) = 2^4 (2\pi)^{-(4s+3k-4)}\Gamma\left(s + \frac{k}{2}\right)\Gamma\left(s + \frac{k}{2} - 1\right)^2\Gamma\left(s + \frac{3k}{2}\right).$$

Exercise 9.7 *Show that the critical points in this case are all integers in the interval* $[-\frac{k}{2} + 2, \frac{k}{2} - 1]$.

To obtain algebraicity results for the special values of $L(s, \pi_F \times \tau_\Psi)$, we realize the integral $Z(s, f, \phi)$ as a Petersson inner product of two Siegel modular forms of degree 2. We will choose $\phi = \phi_F$, which is clearly associated to the Siegel modular form F. The Eisenstein series $E(g, s; f)$ can be used to define the following function $\mathcal{E}$ on $\mathcal{H} := \{Z \in M_2(\mathbb{C}) : i({}^t\bar{Z} - Z) \text{ is positive definite}\}$ by

$$\mathcal{E}(Z, s) = \mu(g)^{-k} \det(J(g, i1_2))^k E\left(g, \frac{s}{3} + \frac{l}{6} - \frac{1}{2}; f\right), \tag{9.1}$$

where $g \in GU(2, 2)^+(\mathbb{R})$ is such that $g\langle i1_2\rangle = Z$. The series that defines $\mathcal{E}(Z, s)$ is absolutely convergent for $\mathrm{Re}(s) > 3 - k/2$ (see [30, p. 210]). Let us assume that $k > 6$. Now we can set $s = 0$, and obtain a holomorphic Eisenstein series $\mathcal{E}(Z, 0)$ on $\mathcal{H}$. Its restriction to $\mathbb{H}_2$, the Siegel upper half-space, is a modular form of weight k with respect to Γ_2. By [38], we know that the Fourier coefficients of $\mathcal{E}(Z, 0)$ are algebraic. Using methods similar to the ones used in Exercise 6.6, we obtain

$$Z(\frac{k}{6} - \frac{1}{2}, f, \bar{\Phi}_F) = \langle \mathcal{E}(\cdot, 0), F\rangle.$$

For any subring $A \subset \mathbb{C}$ denote by $M_k(\Gamma_2, A)$ the A-submodule of $M_k(\Gamma_2)$ consisting of modular forms all of whose Fourier coefficients are contained in A. For a Hecke eigenform $F \in S_k(\Gamma_2)$, let $\mathbb{Q}(F)$ be the subfield of $\mathbb{C}$ generated by all the Hecke eigenvalues of F. From [35, p. 460], we see that $\mathbb{Q}(F)$ is a totally real number field. It is known that $S_k(\Gamma_2)$ has an orthogonal basis of Hecke eigenforms $\{F_j\}$ such that each $F_j \in S_k(\Gamma_2, \mathbb{Q}(F_j))$. Moreover, if F is a Hecke eigenform such that $F \in S_k(\Gamma_2, \mathbb{Q}(F))$, then we can take $F_1 = F$. For details on this, see Sect. 4.3 of [30].

Assume that $F \in S_k(\Gamma_2, \mathbb{Q}(F))$, and consider the basis of $S_k(\Gamma_2)$ containing F as above. Expanding the restriction of $\mathcal{E}(Z, 0)$ to $\mathbb{H}_2$ in terms of this basis, we get for $Z \in \mathbb{H}_2$,

$$\mathcal{E}(Z, 0) = c_F F(Z) + \sum_{j>1} c_{F_j} F_j(Z).$$

The choice of the basis implies that the coefficient c_F is algebraic. Hence, we get

$$c_F = \frac{\langle \mathcal{E}(Z, 0), F\rangle}{\langle F, F\rangle} \in \bar{\mathbb{Q}}.$$

The zeta integral also has contributions from other L-functions of smaller degree and the algebraicity of their special values is known by results of Shimura. Using this, we get

$$\frac{L(\frac{k}{2} - 1, \pi_F \times \tilde{\tau}_\Psi)}{\pi^{5k-8}\langle F, F\rangle\langle \Psi, \Psi\rangle} \in \bar{\mathbb{Q}}.$$

Recall from Exercise 9.7 that $k/2 - 1$ is the rightmost critical point. Let us mention the various people, and their results related to algebraicity of special L-values in this context.

(i) Furusawa [30] got the above result for $F \in S_k(\Gamma_2)$ and $\Psi \in S_k(\Gamma_1)$ and the rightmost critical point.
(ii) Böcherer and Heim [12] got the special value result for $F \in S_k(\Gamma_2)$ and $\Psi \in S_l(\Gamma_1)$ such that k, l are even and $k/2 < l/2 < k - 1$. They also got the result for all critical points which are integers in the set $[l, 2k - 3]$. They did not use the integral representation presented above but had a classical version.

(iii) Pitale, Saha and Schmidt (in various combinations in [68, 69, 75, 85, 86]) got the special value result for $F \in S_k(B(M))$, where $B(M)$ is the Borel congruence subgroup of square free level M, and $\Psi \in S_k(\Gamma_0(N), \chi')$, for any integer N and nebentypus character χ'. They got the result for the critical values which are integers in $[2, k/2 - 1]$. This used theory of differential operators and nearly holomorphic Siegel modular forms.

(iv) Morimoto [63] and Furusawa-Morimoto [31] have obtained special value results in the most general setting. They allow π to be any cuspidal automorphic representation of GSp$_4(\mathbb{A})$ with π_∞ being in the holomorphic discrete series. The representation τ corresponds to $\Psi \in S_k(\Gamma_0(N), \chi')$, where the weight k matches the parameter for π_∞.

Chapter 10
Integral Representation of the Standard L-Function

In this chapter, we want to discuss an integral representation for the standard L-function of degree n holomorphic Siegel cusp forms twisted by characters. The goal is to assume as little as possible about the Siegel modular forms. We will allow vector-valued modular forms with respect to any congruence subgroup. Rallis, Piatetski-Shapiro, and also Böcherer, have obtained the integral representation in special cases. We will modify their methods to suit our level of generality. We will illustrate an application of the integral representation to algebraicity of special values of the L-function in the genus 2 case. We will apply this arithmetic result to the special case of the Siegel modular forms, constructed by Ramakrishnan and Shahidi [80], whose spin L-function is equal to the symmetric cube L-function of elliptic cusp forms. This will give us algebraicity of the symmetric fourth power L-function of an elliptic cusp form twisted by characters. The main reference for this chapter is [72].

10.1 The zeta integral

Since we have to work with symplectic groups of different sizes, let us add a subscript and denote by G_{2n} the group GSp_{2n}. Let P_{2n} be the Siegel parabolic subgroup of G_{2n}, consisting of matrices whose lower left $n \times n$-block is zero. Let $\delta_{P_{2n}}$ be the modulus character of $P_{2n}(\mathbb{A})$. It is given by

$$\delta_{P_{2n}}\left(\begin{bmatrix} A & X \\ & v\,{}^tA^{-1} \end{bmatrix}\right) = |v^{-\frac{n(n+1)}{2}}\det(A)^{n+1}|, \qquad \text{where } A \in \mathrm{GL}_n(\mathbb{A}),\ v \in \mathrm{GL}_1(\mathbb{A}),$$

and $|\cdot|$ denotes the global absolute value, normalized in the standard way.

© Springer Nature Switzerland AG 2019
A. Pitale, *Siegel Modular Forms*, Lecture Notes in Mathematics 2240,
https://doi.org/10.1007/978-3-030-15675-6_10

Exercise 10.1 *Let* $H_{2a,2b} := \{(g, g') \in G_{2a} \times G_{2b} : \mu_a(g) = \mu_b(g')\}$. *Show that the map*

$$H_{2a,2b} \ni \left(\begin{bmatrix} A_1 & B_1 \\ C_1 & D_1 \end{bmatrix}, \begin{bmatrix} A_2 & B_2 \\ C_2 & D_2 \end{bmatrix} \right) \longmapsto \begin{bmatrix} A_1 & & -B_1 & \\ & A_2 & & B_2 \\ -C_1 & & D_1 & \\ & C_2 & & D_2 \end{bmatrix} \in GSp_{2a+2b}$$

is an embedding of $H_{2a,2b}$ *to* G_{2a+2b}.

We will also let $H_{2a,2b}$ denote its image in G_{2a+2b}. Let F be a number field. Let χ be a character of $F^\times \backslash \mathbb{A}^\times$. We define a character on $P_{4n}(\mathbb{A})$, also denoted by χ, by $\chi(p) = \chi(d(p))$. Here, $d(p) = v^{-n} \det(A)$ for $p = \begin{bmatrix} A & \\ & v\,{}^tA^{-1} \end{bmatrix} n'$ with $v \in GL_1(\mathbb{A})$, $A \in GL_{2n}(\mathbb{A})$ and $n' \in N_{4n}(\mathbb{A})$, the unipotent radical of $P_{4n}(\mathbb{A})$. For a complex number s, let

$$I(\chi, s) = \text{Ind}_{P_{4n}(\mathbb{A})}^{G_{4n}(\mathbb{A})} (\chi \delta_{P_{4n}}^s).$$

Thus, $f(\cdot, s) \in I(\chi, s)$ is a smooth $\mathbb{C}$-valued function on $G_{4n}(\mathbb{A})$ satisfying

$$f(pg, s) = \chi(p) \delta_{P_{4n}}(p)^{s+\frac{1}{2}} f(g, s)$$

for all $p \in P_{4n}(\mathbb{A})$ and $g \in G_{4n}(\mathbb{A})$. Consider the Eisenstein series on $G_{4n}(\mathbb{A})$ which, for $\Re(s) > \frac{1}{2}$, is given by the absolutely convergent series

$$E(g, s, f) = \sum_{\gamma \in P_{4n}(F) \backslash G_{4n}(F)} f(\gamma g, s), \tag{10.1}$$

and defined by meromorphic continuation outside this region. Let π be a cuspidal automorphic representation of $G_{2n}(\mathbb{A})$. Let V_π be the space of cuspidal automorphic forms realizing π. For any automorphic form ϕ in V_π, and any $s \in \mathbb{C}$, define a function $Z(s; f, \phi)$ on $G_{2n}(F) \backslash G_{2n}(\mathbb{A})$ by

$$Z(s; f, \phi)(g) = \int_{Sp_{2n}(F) \backslash g \cdot Sp_{2n}(\mathbb{A})} E((h, g), s, f) \phi(h) \, dh. \tag{10.2}$$

10.2   The Basic Identity

We want to unwind the integral $Z(s; f, \phi)$ by substituting the definition of the Eisenstein series. By Proposition 2.4 of [97], we have the double coset decomposition

$$G_{4n}(F) = \bigsqcup_{r=0}^{n} P_{4n}(F) Q_r H_{2n,2n}(F).$$

Here

$$Q_r = \begin{bmatrix} 1_n & 0 & 0 & 0 \\ 0 & I'_{n-r} & 0 & \tilde{I}_r \\ 0 & 0 & 1_n & \tilde{I}_r \\ \tilde{I}_r & -\tilde{I}_r & 0 & I'_{n-r} \end{bmatrix},$$

where the $n \times n$ matrix $\tilde{I}_r$ is given by $\tilde{I}_r = \begin{bmatrix} 0_{n-r} & 0 \\ 0 & 1_r \end{bmatrix}$ and $I'_{n-r} = 1_n - \tilde{I}_r = \begin{bmatrix} 1_{n-r} & 0 \\ 0 & 0 \end{bmatrix}$.

Exercise 10.2 *For* $0 \le r \le n$, *suppose we have the disjoint single coset decomposition*

$$P_{4n}(F)Q_r H_{2n,2n}(F) = \bigsqcup_i P_{4n}(F)Q_r \gamma_i^{(r)}.$$

Show that

$$Z(s; f, \phi) = \sum_{r=0}^{n} Z_r(s; f, \phi),$$

where

$$Z_r(s; f, \phi)(g) := \int_{\mathrm{Sp}_{2n}(F) \backslash g \cdot \mathrm{Sp}_{2n}(\mathbb{A})} \sum_i f(Q_r \gamma_i^{(r)}(h, g), s) \phi(h) \, dh.$$

If $r < n$, then we can do an inner unipotent integral. Lemma 2.2 of [72] states that the section f is invariant under the unipotent. This gives us an integral of the cusp form ϕ over an unipotent subgroup. The cuspidality of ϕ then gives us that $Z_r = 0$ for $0 \le r < n$. For $r = n$, we have

$$P_{4n}(F)Q_n H_{2n,2n}(F) = \bigsqcup_{x \in \mathrm{Sp}_{2n}(F)} P_{4n}(F)Q_n(x, 1).$$

Hence,

$$\begin{aligned} Z_n(s; f, \phi)(g) &= \int_{\mathrm{Sp}_{2n}(F) \backslash g \cdot \mathrm{Sp}_{2n}(\mathbb{A})} \sum_{x \in \mathrm{Sp}_{2n}(F)} f(Q_n \cdot (xh, g), s) \phi(h) \, dh \\ &= \int_{g \cdot \mathrm{Sp}_{2n}(\mathbb{A})} f(Q_n \cdot (h, g), s) \phi(h) \, dh \\ &= \int_{\mathrm{Sp}_{2n}(\mathbb{A})} f(Q_n \cdot (gh, g), s) \phi(gh) \, dh \\ &= \int_{\mathrm{Sp}_{2n}(\mathbb{A})} f(Q_n \cdot (h, 1), s) \phi(gh) \, dh, \end{aligned}$$

For the last equality, we again use Lemma 2.2 of [72]. This gives us

$$Z(s; f, \phi)(g) = \int_{\mathrm{Sp}_{2n}(\mathbb{A})} f(Q_n \cdot (h, 1), s) \phi(gh)\, dh.$$

Theorem 10.3 (Basic identity) *Let* $\phi \in V_\pi$ *be a cusp form which corresponds to a pure tensor* $\otimes_v \phi_v$ *via the isomorphism* $\pi \cong \otimes \pi_v$. *Assume that* $f \in I(\chi, s)$ *factors as* $\otimes f_v$ *with* $f_v \in I(\chi_v, s)$. *Let the function* $Z(s; f, \phi)$ *on* $G_{2n}(F) \backslash G_{2n}(\mathbb{A})$ *be defined as in* (10.2). *Then,* $Z(s; f, \phi)$ *also belongs to* V_π *and corresponds to the pure tensor* $\otimes_v Z_v(s; f_v, \phi_v)$, *where*

$$Z_v(s; f_v, \phi_v) := \int_{\mathrm{Sp}_{2n}(F_v)} f_v(Q_n \cdot (h, 1), s) \pi_v(h) \phi_v\, dh \in \pi_v.$$

10.3   The Local Integral Computation

The goal is now to choose local functions f_v and ϕ_v, and compute the local integrals Z_v defined above. For brevity of notation, let us drop the subscript v. So, let F be a local field. In the non-archimedean case, denote by $\mathfrak{o}, \mathfrak{p}, \varpi, q$ the ring of integers of F, the prime ideal in $\mathfrak{o}$, a uniformizer in $\mathfrak{o}$, and size of the residue field, respectively.

The unramified computation: Assume that χ and π are unramified. Choose $f \in I(\chi, s)$ to be the unramified vector. Hence $f : G_{4n}(F) \to \mathbb{C}$ is given by

$$f\left(\begin{bmatrix} A & * \\ & u\,{}^t A^{-1} \end{bmatrix} k\right) = \chi(u^{-n} \det(A)) \left| u^{-n} \det(A) \right|^{(2n+1)(s+1/2)}$$

for $A \in \mathrm{GL}_{2n}(F)$, $u \in F^\times$ and $k \in G_{4n}(\mathfrak{o})$. Let v_0 be a spherical vector in π. The computation of the local integral in this case makes use of the action of the Hecke algebra and the Satake isomorphism. Here, we use methods similar to [7] and [66], with suitable modifications to incorporate characters. We get

$$Z(s; f, v_0) = \frac{L((2n+1)s + 1/2, \pi \boxtimes \chi, \varrho_{2n+1})}{L((2n+1)(s+1/2), \chi) \prod_{i=1}^{n} L((2n+1)(2s+1) - 2i, \chi^2)}\, v_0$$

(10.3)

for real part of s large enough. See Proposition 4.1 of [72]. Here, the L-function is defined as

$$L(s, \pi \boxtimes \chi, \varrho_{2n+1}) = \frac{1}{1 - \chi(\varpi)q^{-s}} \prod_{i=1}^{n} \frac{1}{(1 - \chi(\varpi)\alpha_i q^{-s})(1 - \chi(\varpi)\alpha_i^{-1} q^{-s})},$$

(10.4)

where α_i are the Satake parameters of π.

The ramified computation: For genus greater than 2, we do not have the local Langlands correspondence for GSp_{2n}. Hence, in the ramified case, we do not even know the appropriate definition of the L-function. All we require out of the local integral in the ramified case is that we make choices of f and v such that $Z(s; f, v)$ is a nonzero multiple of v. With this in mind, let m be a positive integer such that $\chi|_{(1+\mathfrak{p}^m)\cap\mathfrak{o}^\times} = 1$, and such that there exists a vector ϕ in π fixed by $\Gamma_{2n}(\mathfrak{p}^m)$. Here

$$\Gamma_{2n}(\mathfrak{p}^m) = \{g \in Sp_{2n}(\mathfrak{o}) : g \equiv 1_{2n} \bmod \mathfrak{p}^m\}.$$

Let $f(g, s)$ be the unique function on $G_{4n}(F) \times \mathbb{C}$ such that

(1) $f(pg, s) = \chi(p)\delta_{P_{4n}}(p)^{s+\frac{1}{2}} f(g, s)$ for all $g \in G_{4n}(F)$ and $p \in P_{4n}(F)$.
(2) $f(gk, s) = f(g, s)$ for all $g \in G_{4n}(F)$ and $k \in \Gamma_{4n}(\mathfrak{p}^m)$.
(3) $f(Q_n, s) = 1$.
(4) $f(g, s) = 0$ if $g \notin P_{4n}(F)Q_n\Gamma_{4n}(\mathfrak{p}^m)$.

Exercise 10.4 *It is shown in Lemma 4.2 of [72] that the following holds: Let m be a positive integer. Let $p \in P_{4n}(F)$ and $h \in Sp_{2n}(F)$ be such that $Q_n^{-1} p\, Q_n(h, 1) \in \Gamma_{4n}(\mathfrak{p}^m)$. Then $h \in \Gamma_{2n}(\mathfrak{p}^m)$ and $p \in P_{4n}(F) \cap \Gamma_{4n}(\mathfrak{p}^m)$. Use this to show that f is well defined.*

With these choices of f and v, in Proposition 4.3 of [72], we compute

$$Z(s; f, v) = \mathrm{vol}(\Gamma_{2n}(\mathfrak{p}^m))v. \tag{10.5}$$

The archimedean computation: This is by far the most complicated computation. So far, we have not assumed anything about the field F or π and χ. For the archimedean case, assume that F is totally real, and π is in the holomorphic discrete series of $G_{2n}(\mathbb{R})$. These are parametrized by $(k_1, \cdots, k_n)$, where $k_i \in \mathbb{Z}$ are integers all of the same parity, and $k_1 \geq k_2 \geq \cdots \geq k_n > n$. Set $k = k_1$. If all the k_i's are equal, then π corresponds to a weight k scalar valued holomorphic Siegel modular form like the one considered throughout this book. Otherwise, we are in the case of a vector-valued holomorphic Siegel modular form. We also assume that $\chi = \mathrm{sgn}^k$. The induced representation $I(\chi, s)$ has a unique (up to scalar) vector f_k with weight $(k, \cdots, k)$. Choose $f = f_k$. We can show that π also contains a unique (up to scalar) vector v_0 with scalar weight $(k, \cdots, k)$. Then Proposition 5.8 of [72] gives us

$$Z(s, f_k, v_0) = i^{nk} \pi^{n(n+1)/2} A_{\mathbf{k}}((2n + 1)s - 1/2) v_0 \tag{10.6}$$

where the function $A_{\mathbf{k}}(z)$ is defined as

$$A_{\mathbf{k}}(z) = 2^{-n(z-1)}\left(\prod_{j=1}^{n}\prod_{i=1}^{j} \frac{1}{z + k - 1 - j + 2i}\right)\left(\prod_{j=1}^{n}\prod_{i=0}^{\frac{k-k_j}{2}-1} \frac{z - (k - 1 - j - 2i)}{z + (k - 1 - j - 2i)}\right).$$

Recall that we want to choose f and v such that the integral Z is nonzero. Clearly, there are values of s for which the integral is 0 or not defined. For arithmetic applications, it turns out that we need the nonvanishing only for certain critical points. This is checked in the next exercise.

Exercise 10.5 *For integers t such that $0 \leq t \leq k_n - n$, show that $A_{\mathbf{k}}(t)$ is a nonzero rational number.*

10.4   Global Integral Representation

Consider the global field $F = \mathbb{Q}$ and its ring of adeles $\mathbb{A} = \mathbb{A}_{\mathbb{Q}}$. All the results are easily generalizable to a totally real number field. Let $\pi \cong \otimes \pi_p$ be a cuspidal automorphic representation of $G_{2n}(\mathbb{A})$. We assume that π_{∞} is in the holomorphic discrete series representation $\pi_{\mathbf{k}}$ with $\mathbf{k} = k_1 e_1 + \ldots + k_n e_n$, where $k_1 \geq \ldots \geq k_n > n$, and all k_i have the same parity. We set $k = k_1$. Let $\chi = \otimes \chi_p$ be a character of $\mathbb{Q}^{\times} \backslash \mathbb{A}^{\times}$ such that $\chi_{\infty} = \mathrm{sgn}^k$. Let $N = \prod_{p|N} p^{m_p}$ be an integer such that

- For each finite prime $p \nmid N$ both π_p and χ_p are unramified.
- For a prime $p|N$, we have $\chi_p|_{(1+p^{m_p}\mathbb{Z}_p) \cap \mathbb{Z}_p^{\times}} = 1$, and π_p has a vector ϕ_p that is right invariant under the principal congruence subgroup $\Gamma_{2n}(p^{m_p})$ of $\mathrm{Sp}_{2n}(\mathbb{Z}_p)$.

Let ϕ be a cusp form in the space of π corresponding to a pure tensor $\otimes \phi_p$, where the local vectors are chosen as follows. For $p \nmid N$ choose ϕ_p to be a spherical vector; for $p|N$ choose ϕ_p to be a vector right invariant under $\Gamma_{2n}(p^{m_p})$; and for $p = \infty$ choose ϕ_{∞} to be a vector in π_{∞} with scalar weight k. Let $f = \otimes f_p \in I(\chi, s)$ be composed of the following local sections. For a finite prime $p \nmid N$, let f_p be the spherical vector normalized by $f_p(1) = 1$; for $p|N$ choose f_p as in the previous subsection (with the positive integer m of that section equal to the m_p above); and for $p = \infty$, choose f_{∞} to be the scalar weight k vector in $I(\chi_{\infty}, s)$. Define $L^N(s, \pi \boxtimes \chi, \varrho_{2n+1}) = \prod_{\substack{p \nmid N \\ p \neq \infty}} L(s, \pi_p \boxtimes \chi_p, \varrho_{2n+1})$, where the local factors on the right were defined in (10.4). Using (10.3), (10.5) and (10.6) we obtain the following theorem.

Theorem 10.6 (Pitale et al. [72]) *Let the notation be as above. Then, the function $L^N(s, \pi \boxtimes \chi, \varrho_{2n+1})$ can be analytically continued to a meromorphic function of s with only finitely many poles. Furthermore, for all $s \in \mathbb{C}$ and $g \in G_{2n}(\mathbb{A})$,*

$$Z(s, f, \phi)(g) = \frac{L^N((2n+1)s + 1/2, \pi \boxtimes \chi, \varrho_{2n+1})}{L^N((2n+1)(s+1/2), \chi) \prod_{j=1}^n L^N((2n+1)(2s+1) - 2j, \chi^2)}$$

$$\times i^{nk} \pi^{n(n+1)/2} \left(\prod_{p|N} \mathrm{vol}(\Gamma_{2n}(p^{m_p})) \right) A_{\mathbf{k}}((2n+1)s - 1/2)\phi(g).$$

$$(10.7)$$

10.5 Classical Reformulation

For $\phi \in V_\pi$, define a function $F : \mathbb{H}_n \to \mathbb{C}$ by

$$F(Z) = \det(J(g, I))^k \, \phi(g),$$

where g is any element of $\mathrm{Sp}_{2n}(\mathbb{R})$ with $g\langle I \rangle = Z$. This F transforms like a Siegel modular form of weight k excepting that it will not be holomorphic in the vector-valued case. Define the Eisenstein series on $\mathbb{H}_{2n}$ by

$$E^\chi_{k,N}(Z, s) := \det(J(g, I))^k E\left(g, \frac{2s}{2n+1} + \frac{k}{2n+1} - \frac{1}{2}, f\right),$$

where g is any element of $\mathrm{Sp}_{4n}(\mathbb{R})$ with $g\langle I \rangle = Z$. We know the following about the Eisenstein series (Theorem 17.9 of [98]).

Proposition 10.7 *Suppose that $k \geq 2n + 2$. Then, the series defining $E^\chi_{k,N}(Z, 0)$ is absolutely convergent, and defines a holomorphic Siegel modular form of degree $2n$ and weight k with respect to the principal congruence subgroup $\Gamma_{4n}(N)$ of $\mathrm{Sp}_{4n}(\mathbb{R})$. More generally, let $0 \leq m \leq \frac{k}{2} - n - 1$ be an integer. Then $E^\chi_{k,N}(Z, -m)$ is a nearly holomorphic Siegel modular form of weight k with respect to $\Gamma_{4n}(N)$.*

Restrict $E^\chi_{k,N}$ to $\begin{bmatrix} Z_1 \\ & Z_2 \end{bmatrix}$ with $Z_1, Z_2 \in \mathbb{H}_n$. As in Sect. 9.4, we can rewrite the integral $Z(s, f, \phi)$ as the Petersson inner product of F and the restriction of $E^\chi_{k,N}$. Then Theorem 10.6 translates to

$$\langle E^\chi_{k,N}(-\,,Z_2, \tfrac{n}{2} - \tfrac{k-s}{2}), \bar{F} \rangle = \frac{L^N(s, \pi \boxtimes \chi, \varrho_{2n+1})}{L^N(s+n, \chi) \prod_{j=1}^n L^N(2s+2j-2, \chi^2)} \times A_{\mathbf{k}}(s-1)$$

$$\times \prod_{p|N} \mathrm{vol}(\Gamma_{2n}(p^{m_p})) \times \frac{i^{nk} \pi^{n(n+1)/2}}{\mathrm{vol}(\mathrm{Sp}_{2n}(\mathbb{Z}) \backslash \mathrm{Sp}_{2n}(\mathbb{R}))} \times F(Z_2).$$

10.6 Arithmetic Results in Genus 2

If the Eisenstein series $E^\chi_{k,N}$ is holomorphic, then we know that it has algebraic Fourier coefficients. Similarly, we can find a basis of holomorphic Siegel cusp forms of weight k and genus n with respect to a congruence subgroup such that their Fourier coefficients are algebraic. Unfortunately, in the classical reformulation of the integral representation of the L-function, we are taking the inner product of two Siegel modular forms, neither of them are holomorphic. All we know is that the modular forms are nearly holomorphic. If we knew that any nearly holomorphic Siegel modular form is obtained from holomorphic ones by applying differential operators which preserve the algebraicity of Fourier coefficients, then we could obtain algebraicity of the Petersson inner product and hence the special L-values. Such a structure theorem is only proven for genus two in [71]. Hence, we have the following theorem.

Theorem 10.8 (Pitale et al. [72]) *Let π be a cuspidal automorphic representation of $GSp(4, \mathbb{A}_{\mathbb{Q}})$ such that π_∞ is isomorphic to the holomorphic discrete series representation with highest weight (k_1, k_2) such that $k_1 \geq k_2 \geq 3, k_1 \equiv k_2 \pmod 2$. Let χ be any Dirichlet character satisfying $\chi(-1) = (-1)^{k_2}$. Let r be any integer satisfying $1 \leq r \leq k_2 - 2, r \equiv k_2 \pmod 2$. Furthermore, if $\chi^2 = 1$, we assume that $r \neq 1$. Let $\tau(\chi)$ be the Gauss sum attached to χ. Then*

$$\frac{L(r, \pi \boxtimes \chi, \varrho_5)}{(2\pi i)^{2k+3r} \tau(\chi)^3 \langle F, F \rangle} \in \bar{\mathbb{Q}}.$$

Previous results were those of Shimura's [98] for $k_1 = k_2$ and with respect to $\Gamma_0^{(2)}(N)$, and of Kozima's [53] for $k_1 > k_2$ but only for full level and $\chi = 1$.

Symmetric fourth L-function of GL_2: Let k be an even positive integer and M any positive integer. Let f be an elliptic cuspidal newform of weight k, level M and trivial nebentypus that is not of dihedral type. According to Theorem A' and Theorem C of [80], there exists a cuspidal automorphic representation π of $GSp_4(\mathbb{A})$ such that

(1) π_∞ is the holomorphic discrete series representation with highest weight $(2k - 1, k + 1)$,
(2) for $p \nmid M$, the local representation π_p is unramified,
(3) the L-functions have the following relation.

$$L(s, \pi, \text{spin}) = L(s, \text{sym}^3 f).$$

If α_p, α_p^{-1} are the Satake p-parameters of f for $p \nmid M$, then the local factor of the mth symmetric power L-function of f is the degree $m + 1$ L-function

$$L_p(s, \text{sym}^m f) = \prod_{i=0}^m (1 - \alpha_p^{m-2i} p^{-s})^{-1}.$$

Note that π corresponds to a holomorphic vector-valued Siegel cusp form F_0 with weight $\det^{k+1} \text{sym}^{k-2}$. Let χ be a Dirichlet character with $\chi_\infty = \text{sgn}$.

Exercise 10.9 *Show that*

$$L(s, \pi \boxtimes \chi, \varrho_5) = L(s, \chi \otimes \text{sym}^4 f). \tag{10.8}$$

Here, on the right-hand side, we have the L-function of GL_5 given by the symmetric fourth power of f, twisted by the character χ.

From Theorems 10.8 and (10.8), we get the following theorem.

Theorem 10.10 (Pitale et al. [72]) *Let f, π be as above, and let $F \in \pi$ be such that its Fourier coefficients lie in a CM field. Let S be any finite set of places of $\mathbb{Q}$ containing the infinite place, χ be a Dirichlet character with $\chi_\infty = \mathrm{sgn}$, and r be an integer satisfying $1 \le r \le k - 1$, r odd. If $r = 1$, assume $\chi^2 \ne 1$. Then we have*

$$\frac{L^S(r, \chi \otimes \mathrm{sym}^4 f)}{(2\pi i)^{4k-2+3r} G(\chi)^3 \langle F, F \rangle} \in \bar{\mathbb{Q}}. \tag{10.9}$$

Deligne's conjecture on critical values of motivic L-functions predicts an algebraicity result for the critical values of $L(s, \chi \otimes \mathrm{sym}^m f)$ for each positive integer m. For $m = 1$ this was proved by Shimura [95], for $m = 2$ by Sturm [100], and for $m = 3$ by Garrett–Harris [36]. In the case $m = 4$, and f of full level, Ibukiyama and Katsurada [40] proved a formula for $L(s, \mathrm{sym}^4 f)$ which implies algebraicity. Assuming functoriality, the expected algebraicity result for the critical values of $L(s, \chi \otimes \mathrm{sym}^m f)$ was proved for all odd m by Raghuram [79]. To the best of our knowledge, Theorem 10.10 represents the first advance in the case $m = 4$ for general newforms f.

Appendix A
GL₂ Notes: Classical Modular Forms

Please refer to Chap. 1 of [17] for basics on classical modular forms.

(1) The general linear group is defined by $GL_2(\mathbb{R}) := \{g \in M_2(\mathbb{R}) : \det(g) \neq 0\}$.
The special linear group is $SL_2(\mathbb{R}) = \{g \in GL_2(\mathbb{R}) : \det(g) = 1\}$.

(2) Consider the subgroups of $GL_2(\mathbb{R})$ given by

$$N := \{\begin{bmatrix} 1 & x \\ & 1 \end{bmatrix} : x \in \mathbb{R}\} \text{ and } A := \{\begin{bmatrix} a & \\ & b \end{bmatrix} : a, b \in \mathbb{R}^\times\}.$$

We have the Bruhat decomposition

$$GL_2(\mathbb{R}) = NA \sqcup NA \begin{bmatrix} & 1 \\ -1 & \end{bmatrix} N.$$

(3) The complex upper half-space is

$$\mathbb{H}_1 := \{z = x + iy \in \mathbb{C} : y > 0\}.$$

The group $GL_2(\mathbb{R})^+ := \{g \in GL_2(\mathbb{R}) : \det(g) > 0\}$ acts on $\mathbb{H}_1$ by the formula

$$g\langle z\rangle := \frac{az + b}{cz + d}, \quad z \in \mathbb{H}_1, g = \begin{bmatrix} a & b \\ c & d \end{bmatrix} \in GL_2(\mathbb{R})^+.$$

We have

$$\mathrm{Im}(g\langle z\rangle) = \frac{\mathrm{Im}(z)}{|cz + d|^2},$$

and the element of volume on $\mathbb{H}_1$, invariant under the above action, is

$$ds = \frac{dx\,dy}{y^2}.$$

© Springer Nature Switzerland AG 2019
A. Pitale, *Siegel Modular Forms*, Lecture Notes in Mathematics 2240,
https://doi.org/10.1007/978-3-030-15675-6

(4) Let $\Gamma_1 := SL_2(\mathbb{Z})$. An elliptic modular form of weight k (a positive integer) with respect to Γ_1 is a holomorphic function $f : \mathbb{H}_1 \to \mathbb{C}$ that is bounded in regions $y \geq y_0$ for any $y_0 > 0$, and satisfies

$$f(\frac{az+b}{cz+d}) = (cz+d)^k f(z) \text{ for all } \begin{bmatrix} a & b \\ c & d \end{bmatrix} \in SL_2(\mathbb{Z}).$$

The space of such forms is denoted by $M_k(\Gamma_1)$.

(5) Every $f \in M_k(\Gamma_1)$ has a Fourier expansion

$$f(z) = \sum_{n \geq 0} A(n)e^{2\pi i n z}.$$

(6) $f \in M_k(\Gamma_1)$ is called a cusp form if $A(0) = 0$. The space of cusp forms is denoted by $S_k(\Gamma_1)$.

(7) If $f, g \in M_k(\Gamma_1)$, at least one of which is a cusp form, then we can define the Petersson inner product

$$\langle f, g \rangle := \int_{\Gamma_1 \backslash \mathbb{H}_1} f(z)\overline{g(z)} y^k \frac{dxdy}{y^2}.$$

A fundamental domain for $\Gamma_1 \backslash \mathbb{H}_1$ is given by the set

$$\{z = x + iy \in \mathbb{H}_1 : |x| \leq \frac{1}{2}, |z| \geq 1\}.$$

(8) Eisenstein series of weight $k \geq 4$ are given by

$$e_k(z) := \sum_{\begin{bmatrix} a & b \\ c & d \end{bmatrix} \in \Gamma_\infty \backslash \Gamma_1} \frac{1}{(cz+d)^k}, \qquad \Gamma_\infty := \{\begin{bmatrix} 1 & n \\ & 1 \end{bmatrix} : n \in \mathbb{Z}\}.$$

Then, $e_k(z) \in M_k(\Gamma_1)$. Note that $e_k = 0$ for k odd.

(9) The Ramanujan delta function, given by the formula

$$\Delta(z) := \sum_{n > 0} \tau(n)e^{2\pi i n z} = q \prod_{n \geq 1}(1 - q^n)^{24}, \text{ where } q = e^{2\pi i z},$$

is a cusp form of weight 12.

(10) Let N be any positive integer. A congruence subgroup of level N is defined by

$$\Gamma_0(N) := \{\begin{bmatrix} a & b \\ c & d \end{bmatrix} \in \Gamma_1 : c \equiv 0 \pmod{N}\}.$$

Modular forms and cusp forms of weight k and level N are defined just as for Γ_1, with additional conditions at the cusps, and are denoted by $M_k(\Gamma_0(N))$ and $S_k(\Gamma_0(N))$.

(11) Let $\mathcal{H}$ be the Hecke algebra for Γ_1 and, for every prime number p, let $\mathcal{H}_p$ be the local Hecke algebra at the prime p. We have $\mathcal{H} = \otimes_p \mathcal{H}_p$. Also, $\mathcal{H}_p$ is generated by the two elements $\Gamma_1 \begin{bmatrix} p & \\ & p \end{bmatrix} \Gamma_1$ and $\Gamma_1 \begin{bmatrix} 1 & \\ & p \end{bmatrix} \Gamma_1$.

(12) Let $T(m) = \sum_{\det(g)=m} \Gamma_1 g \Gamma_1$. Then, for $(m, n) = 1$, we have the relation $T(mn) = T(m)T(n)$. We also have the single coset decomposition

$$T(m) = \bigsqcup_{\substack{a,d>0, ad=m \\ b \bmod d}} \Gamma_1 \begin{bmatrix} a & b \\ & d \end{bmatrix}.$$

In particular,

$$\Gamma_1 \begin{bmatrix} 1 & \\ & p \end{bmatrix} \Gamma_1 = \Gamma_1 \begin{bmatrix} p & \\ & 1 \end{bmatrix} \sqcup \bigsqcup_{b \bmod p} \Gamma_1 \begin{bmatrix} 1 & b \\ & p \end{bmatrix}.$$

(13) Let $T(g) = \Gamma_1 g \Gamma_1 = \sqcup_i \Gamma_1 g_i \in \mathcal{H}$. Then, $T(g)$ acts on $M_k(\Gamma_1)$ (and on $S_k(\Gamma_1)$) as follows. Let $f \in M_k(\Gamma_1)$. Then $T(g)f := \sum_i f|_k g_i$, where the slash $|_k$ action is given by

$$(f|_k g)(z) := \det(g)^{k-1}(cz + d)^{-k} f(g\langle z \rangle), \text{ for } g = \begin{bmatrix} a & b \\ c & d \end{bmatrix} \in GL_2(\mathbb{R})^+, z \in \mathbb{H}_1.$$

(14) Let $\{A(n)\}$ be the Fourier coefficients of $f \in M_k(\Gamma_1)$. Let $\{B(n)\}$ be the Fourier coefficients of $T(m)f$. Then, we have

$$B(n) = \sum_{ad=m, a|n} \left(\frac{a}{d}\right)^{k/2} d A\left(\frac{nd}{a}\right).$$

(15) The space $M_k(\Gamma_1)$ has a basis of simultaneous eigenfunctions of the Hecke algebra $\mathcal{H}$. Furthermore, $S_k(\Gamma_1)$ has such a basis which is orthogonal with respect to the Petersson inner product. A cusp form $S_k(\Gamma_1)$ is called *normalized* if it satisfies $A(1) = 1$. The space $S_k(\Gamma_1)$ has a basis of normalized Hecke eigenforms.

(16) The L-function of a modular form $f \in M_k(\Gamma_1)$, with Fourier coefficients $\{A(n)\}$, is defined by

$$L(s, f) := \sum_{n \geq 1} \frac{A(n)}{n^s}.$$

We have the following properties of the L-function.

(a) The Mellin transform of f gives the completed L-function.

$$\int_0^\infty f(iy)y^s \frac{dy}{y} = (2\pi)^{-s}\Gamma(s)L(s, f) =: \Lambda(s, f).$$

(b) The completed L-function $\Lambda(s, f)$ extends to an analytic function of s if f is a cusp form; if f is not cuspidal, then it has a simple pole at $s = 0$ and $s = k$. It satisfies the functional equation

$$\Lambda(s, f) = (-1)^{k/2}\Lambda(k - s, f).$$

(c) If f is a normalized Hecke eigenform then

$$L(s, f) = \prod_{p \text{ prime}} \left(\sum_{r=0}^\infty A(p^r)p^{-rs}\right) = \prod_{p \text{ prime}} (1 - A(p)p^{-s} + p^{k-1+2s})^{-1}.$$

(17) Every $f \in S_k(\Gamma_0(N))$ with Fourier coefficients $\{A(n)\}$ satisfies the Ramanujan conjecture

$$|A(n)| \le Cn^{\frac{k-1}{2}+\epsilon}, \text{ for all } \epsilon > 0.$$

(18) If f is a Hecke eigenform and $1 - A(p)p^{-s} + p^{k-1+2s} = (1 - \alpha_{0,p}p^{-s})(1 - \alpha_{0,p}\alpha_{1,p}p^{-s})$, then the Ramanujan conjecture states that $|\alpha_{0,p}| = p^{(k-1)/2}$ and $|\alpha_{1,p}| = 1$.

Appendix B
The p-Adic Fields $\mathbb{Q}_p$ and the Ring of Adeles $\mathbb{A}$ of $\mathbb{Q}$

(1) Let p be a prime number. Define $\mathrm{ord}_p : \mathbb{Z}\backslash\{0\} \to \mathbb{N}$ by $\mathrm{ord}_p(x) = \max\{r \in \mathbb{N} : p^r | x\}$. Extend it to $\mathbb{Q}\backslash\{0\}$ by $\mathrm{ord}_p(a/b) = \mathrm{ord}_p(a) - \mathrm{ord}_p(b)$, for $a, b \in \mathbb{Z}$.

(2) The p-adic absolute value $| \ |_p$ on $\mathbb{Q}$ is defined by

$$|x|_p = \begin{cases} p^{-\mathrm{ord}_p(x)} & \text{if } x \neq 0; \\ 0 & \text{if } x = 0. \end{cases}$$

The p-adic absolute value satisfies the stronger condition

$$|x + y|_p \leq \max(|x|_p, |y|_p), \text{ with equality if and only if } |x|_p \neq |y|_p.$$

(3) The completion of $\mathbb{Q}$ with respect to the p-adic absolute value $| \ |_p$ is called the field $\mathbb{Q}_p$ of p-adic numbers. Let $\mathbb{Z}_p := \{\alpha \in \mathbb{Q}_p : |\alpha|_p \leq 1\}$ be the ring of p-adic integers. $\mathbb{Z}_p$ is the completion of $\mathbb{Z}$ in $\mathbb{Q}_p$.

(4) Every p-adic number $\alpha \in \mathbb{Q}_p$ has a unique p-adic expansion of the form

$$\alpha = \alpha_{-r}p^{-r} + \alpha_{-r+1}p^{-r+1} + \cdots + \alpha_{-1}p^{-1} + \alpha_0 + \alpha_1 p + \alpha_2 p^2 + \cdots,$$

with $\alpha_n \in \mathbb{Z}$ such that $0 \leq \alpha_n \leq p - 1$. Furthermore, $\alpha \in \mathbb{Z}_p$ if and only if $\alpha_n = 0$ for all $n < 0$. For α given by its p-adic expansion, the p-adic absolute value of α is given by $|\alpha|_p = p^{-n}$, where n is the least integer such that $\alpha_n \neq 0$.

(5) $\mathbb{Z}_p$ has a unique maximal ideal $p\mathbb{Z}_p$. We can characterize $p\mathbb{Z}_p = \{\alpha \in \mathbb{Z}_p : |\alpha|_p < 1\} = \{\alpha \in \mathbb{Z}_p : \alpha_0 = 0\}$. The group of units in $\mathbb{Z}_p$ is denoted by $\mathbb{Z}_p^\times$. Once again, we can characterize $\mathbb{Z}_p^\times = \{\alpha \in \mathbb{Z}_p : |\alpha|_p = 1\} = \{\alpha \in \mathbb{Z}_p : \alpha_0 \neq 0\}$. The ring $\mathbb{Z}_p$ is compact in $\mathbb{Q}_p$.

(6) We have $\mathbb{Z}_p/p\mathbb{Z}_p \simeq \mathbb{Z}/p\mathbb{Z}$ and $\mathbb{Z}_p^\times/(1 + p\mathbb{Z}_p) \simeq (\mathbb{Z}/p\mathbb{Z})^\times$. Also $\mathbb{Q}_p^\times = \langle p \rangle \times \mathbb{Z}_p^\times$.

(7) We will let $p = \infty$ denote the Archimedean place. We have the ring of direct product of all completions

© Springer Nature Switzerland AG 2019
A. Pitale, *Siegel Modular Forms*, Lecture Notes in Mathematics 2240,
https://doi.org/10.1007/978-3-030-15675-6

$$\prod_{p \leq \infty} \mathbb{Q}_p = \mathbb{R} \times \mathbb{Q}_2 \times \mathbb{Q}_3 \times \mathbb{Q}_5 \times \cdots.$$

The ring of adeles $\mathbb{A}$ is the subring of the full direct product given by

$$\mathbb{A} = \{a = (a_p)_{p \leq \infty} \in \prod_{p \leq \infty} \mathbb{Q}_p : a_p \in \mathbb{Z}_p \text{ for all but finitely many } p\}.$$

We call $\mathbb{A}$ the *restricted direct product* of the $\mathbb{Q}_p$ (with respect to the open-compact subsets $\mathbb{Z}_p$ for $p < \infty$).

(8) We define the topology on $\mathbb{A}$ to be the one generated by the sets $\prod_{p \leq \infty} U_p$, where U_p are open in $\mathbb{Q}_p$ and $U_p = \mathbb{Z}_p$ for almost all p. With this topology, $\mathbb{A}$ is a locally compact topological ring.

(9) We have

$$\mathbb{A}/\mathbb{Q} \simeq \mathbb{R}/\mathbb{Z} \times \prod_{p < \infty} \mathbb{Z}_p.$$

This tells us that $\mathbb{A}/\mathbb{Q}$ is compact.

Appendix C
GL$_2$ Notes: Representation Theory

Please refer to Chaps. 3 and 4 of [17] for the global or local representation theory of GL$_2$.

(1) Strong approximation for SL$_2$ gives us

$$\mathrm{GL}_2(\mathbb{A}) = \mathrm{GL}_2(\mathbb{Q})\mathrm{GL}_2(\mathbb{R})^+ K(N),$$

where

$$K(N) = \prod_{p < \infty} K_p(p^{\mathrm{ord}_p(N)}), \qquad K_p(p^m) := \{\begin{bmatrix} a & b \\ c & d \end{bmatrix} \in \mathrm{GL}_2(\mathbb{Z}_p) : c \in p^m \mathbb{Z}_p\}.$$

(2) Let $f \in S_k(\Gamma_0(N))$. Write $g \in \mathrm{GL}_2(\mathbb{A})$ as $g = \gamma g_\infty k$, with $\gamma \in \mathrm{GL}_2(\mathbb{Q})$, $g_\infty \in \mathrm{GL}_2(\mathbb{R})^+$, $k \in K(N)$. Define

$$\Psi_f(g) := (ci + d)^{-k} f\left(\frac{ai + b}{ci + d}\right), \qquad g_\infty = \begin{bmatrix} a & b \\ c & d \end{bmatrix}.$$

(3) Let (τ_f, V_τ) be the representation of GL$_2(\mathbb{A})$ generated by right translates of Ψ_f. The group acts on V_τ by right translation. If f is a Hecke eigenform, then τ_f is irreducible, and we have the restricted tensor product $\tau_f \cong \otimes_{p \leq \infty} \tau_p$. Here, τ_p is an irreducible, admissible representation of GL$_2(\mathbb{Q}_p)$ (or GL$_2(\mathbb{R})$).

(4) Let p be a finite prime. There are three possibilities for an infinite-dimensional, irreducible, admissible representation of GL$_2(\mathbb{Q}_p)$.

   (a) *Irreducible principal series*: Suppose χ_1, χ_2 are two characters of $\mathbb{Q}_p^\times$ such that $\chi_1 \chi_2^{-1} \neq | \cdot |_p^{\pm 1}$. The *irreducible* representation $\chi_1 \times \chi_2$ is obtained by normalized induction from the character of the Borel subgroup $B(\mathbb{Q}_p)$ of GL$_2(\mathbb{Q}_p)$ obtained from χ_1 and χ_2. The standard model for $\chi_1 \times \chi_2$ is the space of locally constant $\mathbb{C}$-valued functions ϕ on GL$_2(\mathbb{Q}_p)$ satisfying

© Springer Nature Switzerland AG 2019
A. Pitale, *Siegel Modular Forms*, Lecture Notes in Mathematics 2240,
https://doi.org/10.1007/978-3-030-15675-6

$$\phi(\begin{bmatrix} a & b \\ & d \end{bmatrix} g) = |ad^{-1}|^{1/2}\chi_1(a)\chi_2(d)\phi(g).$$

(b) *Twist of Steinberg representation*: Suppose χ is a character of $\mathbb{Q}_p^\times$. The representation $||_p^{1/2}\chi \times ||_p^{-1/2}\chi$ is reducible, and has a unique irreducible subrepresentation called the twist of the Steinberg representation of $GL_2(\mathbb{Q}_p)$. It is denoted by χSt_{GL_2}.

(c) *Supercuspidal representations*: Any irreducible representation of $GL_2(\mathbb{Q}_p)$ that cannot be obtained as a subrepresentation of any representation induced from the Borel subgroup is called a supercuspidal representation. One can construct these by induction from compact subgroups of $GL_2(\mathbb{Q}_p)$.

(5) If $p = \infty$, the possibilities for an irreducible, admissible representation of $GL_2(\mathbb{R})$ are either the discrete series $D_\mu(\ell)$, $\ell \geq 1$ or a limit of discrete series representation $D_\mu(0)$ or a principle series representation. In the discrete series case, $\mu \in \mathbb{C}$ gives the central character, and $\ell + 1$ gives the lowest weight of the representation.

(6) Back to $f \in S_k(\Gamma_0(N))$ and $\tau_f \cong \otimes_p \tau_p$. We can describe τ_p as follows.

(a) Let p be a prime such that p does not divide N. Then $\tau_p = \chi_p \times \chi_p^{-1}$, where the character χ_p is unramified, i.e., trivial on $\mathbb{Z}_p^\times$. If λ_p is the pth-Hecke eigenvalue of f, then we have

$$\lambda_p = p^{\frac{k-1}{2}}(\chi_p(p) + \chi_p^{-1}(p)).$$

(b) Let p be a prime such that $p|N$ but p^2 does not divide N. Then $\tau_p = \chi_p St_{GL_2}$ for an unramified character χ_p of $\mathbb{Q}_p^\times$. Suppose f is an eigenfunction of the Atkin–Lehner operator with eigenvalue ϵ_p. Then, $\chi_p(p) = -\epsilon_p$.

(c) Let p be a prime such that $p^2|N$. Then, τ_p could be any of the three types of representations mentioned above.

(d) At the archimedean place, τ_∞ is the discrete series representation $D_0(k-1)$.

(7) The L-function for the non-archimedean representations is as follows:

$$L(s, \tau_p) = \begin{cases} L(s, \chi_1)L(s, \chi_2) & \text{if } \tau_p = \chi_1 \times \chi_2; \\ L(s - 1/2, \chi) & \text{if } \tau_p = \chi St_{GL_2}; \\ 1 & \text{otherwise.} \end{cases}$$

Recall that

$$L(s, \chi) = \begin{cases} (1 - \chi(p)p^{-s})^{-1} & \text{if } \chi \text{ is unramified;} \\ 1 & \text{if } \chi \text{ is ramified.} \end{cases}$$

(8) *Newforms theory for* GL$_2$: Let (τ, V) be an irreducible admissible infinite-dimensional representation of GL$_2(\mathbb{Q}_p)$. For $n \in \mathbb{Z}, n \geq 0$, set $V^{(n)} := \{v \in V : \tau(g)v = v,$ for all $g \in K_p(p^n)\}$. It is known that there is a n such that $V^{(n)} \neq 0$. Suppose n_0 is the least n such that $V^{(n)} \neq 0$. Then, $\dim(V^{(n_0)}) = 1$. Denote $a(\tau) = n_0$ and we call $p^{a(\tau)}\mathbb{Z}_p$ the conductor of τ. If χ is a character of $\mathbb{Q}_p^\times$, then $a(\chi)$ denotes the smallest nonnegative integer m such that χ is trivial on $(1 + p^m\mathbb{Z}_p) \cap \mathbb{Z}_p^\times$. Then, we have the following:

$$a(\chi_1 \times \chi_2) = a(\chi_1) + a(\chi_2),$$

$$a(\chi\text{St}_{\text{GL}_2}) = \begin{cases} 1 & \text{if } a(\chi) = 0; \\ 2a(\chi) & \text{if } a(\chi) > 0. \end{cases}$$

(9) *Kirillov model*: Choose a nontrivial additive character ψ of $\mathbb{Q}_p$. It is known that for every irreducible admissible infinite-dimensional representation τ of GL$_2(\mathbb{Q}_p)$, there is a unique space $\mathcal{K}(\tau, \psi)$ of locally constant functions $\phi : \mathbb{Q}_p^\times \to \mathbb{C}$ with the following property: GL$_2(\mathbb{Q}_p)$ acts on $\mathcal{K}(\tau, \psi)$ in a way such that

$$\left(\begin{bmatrix} a & b \\ & d \end{bmatrix}\phi\right)(x) = \omega_\tau(d)\psi(bx/d)\phi(ax/d),$$

and the resulting representation of GL$_2(\mathbb{Q}_p)$ is equivalent to τ (here ω_τ is the central character of τ). This is called the Kirillov model of τ (with respect to ψ).

(10) *Newforms in Kirillov model*: Let ϕ be a newform (unique up to scalars) in a Kirillov model of τ, i.e., $\phi \in V_\tau^{(n_0)}$. Let 1_U be the characteristic function of a subset U of $\mathbb{Q}_p^\times$. Then, the formula for ϕ is given by

Representation	Local new form $\phi(x)$
$\chi_1 \times \chi_2, \chi_i$ unramified	$|x|_p^{1/2}\left(\sum_{k+l=\text{ord}_p(x)} \chi_1(p^k)\chi_2(p^l)\right)1_{\mathbb{Z}_p}(x)$
$\chi_1 \times \chi_2, \chi_1$ ram., χ_2 unram.	$|x|_p^{1/2}\chi_2(x)1_{\mathbb{Z}_p}(x)$
$\chi_1 \times \chi_2, \chi_1$ unram., χ_2 ram.	$|x|_p^{1/2}\chi_1(x)1_{\mathbb{Z}_p}(x)$
$\chi_1 \times \chi_2, \chi_i$ ramified	$1_{\mathbb{Z}_p^\times}(x)$
$\chi\text{St}_{\text{GL}_2}, \chi$ unram.	$|x|_p\chi(x)1_{\mathbb{Z}_p}(x)$
$\chi\text{St}_{\text{GL}_2}, \chi$ ram. or supercuspidal	$1_{\mathbb{Z}_p^\times}(x)$

Appendix D
Solutions to Exercises

Chapter 1

Exercise 1.1: Since A is positive definite, one can diagonalize A over $\mathbb{R}$, and show that $M_t := \{x \in \mathbb{R}^m : Q(x) = t\}$ is compact. Since $M_t \cap \mathbb{Z}^m$ is the intersection of a compact and a discrete set, it is finite.

Exercise 1.2: We have $\Theta_Q = \alpha E_1 + \beta E_2$. Compare the first two coefficients of both sides to conclude

$$1 = -3\alpha - \beta, \qquad 8 = -24\alpha - 24\beta.$$

Solving for α, β gives us $\Theta_Q = -1/3E_1$. Hence,

$$r_Q(n) = 8\big(\sigma_1(n) - 4\sigma_1(\frac{n}{4})\big), \qquad \text{for all } n \geq 1.$$

So, we get $r_Q(n) \geq 1$, whenever $n \geq 1$.

Exercise 1.3: For $g = \begin{bmatrix} A & B \\ C & D \end{bmatrix}$, we have

$$
\begin{aligned}
{}^t\!gJg &= \begin{bmatrix} {}^t\!A & {}^t\!C \\ {}^t\!B & {}^t\!D \end{bmatrix} \begin{bmatrix} & 1_n \\ -1_n & \end{bmatrix} \begin{bmatrix} A & B \\ C & D \end{bmatrix} = \begin{bmatrix} {}^t\!A & {}^t\!C \\ {}^t\!B & {}^t\!D \end{bmatrix} \begin{bmatrix} C & D \\ -A & -B \end{bmatrix} \\
&= \begin{bmatrix} {}^t\!AC - {}^t\!CA & {}^t\!AD - {}^t\!CB \\ {}^t\!BC - {}^t\!DA & {}^t\!BD - {}^t\!DB \end{bmatrix}.
\end{aligned}
$$

This immediately gives us (i) $\Leftrightarrow$ (iv). Next,

$$
{}^t\!gJg = \mu J \Leftrightarrow g^{-1} = \mu^{-1} J^{-1}\, {}^t\!gJ = \mu^{-1} \begin{bmatrix} {}^t\!D & -{}^t\!B \\ -{}^t\!C & {}^t\!A \end{bmatrix}.
$$

© Springer Nature Switzerland AG 2019
A. Pitale, *Siegel Modular Forms*, Lecture Notes in Mathematics 2240,
https://doi.org/10.1007/978-3-030-15675-6

This gives us (i) $\Leftrightarrow$ (iii). Now, we have

$$\mu g g^{-1} = \mu 1_{2n} \Leftrightarrow \begin{bmatrix} A & B \\ C & D \end{bmatrix} \begin{bmatrix} {}^t D & -{}^t B \\ -{}^t C & {}^t A \end{bmatrix} = \mu 1_{2n}$$

$$\Leftrightarrow \begin{bmatrix} A\,{}^t D - B\,{}^t C & B\,{}^t A - A\,{}^t B \\ C\,{}^t D - D\,{}^t C & D\,{}^t A - C\,{}^t B \end{bmatrix} = \mu 1_{2n}.$$

This gives us (iii) $\Leftrightarrow$ (v). Arguing as before, we finally get (ii) $\Leftrightarrow$ (v).

Exercise 1.4: The definition of $\mathrm{Sp}_{2n}(\mathbb{R})$ tells us that, for $g \in \mathrm{Sp}_{2n}(\mathbb{R})$, we have $\det(g) = \pm 1$. It is clear that $\begin{bmatrix} 1_n & X \\ & 1_n \end{bmatrix}$ and $\begin{bmatrix} g & \\ & {}^t g^{-1} \end{bmatrix}$ have determinant 1. So it is enough to show that the determinant of any element of K_n is positive (hence, equal to 1). For that, we have

$$\det(\begin{bmatrix} X & Y \\ -Y & X \end{bmatrix}) = \det(\begin{bmatrix} 1_n & i 1_n \\ & 1_n \end{bmatrix} \begin{bmatrix} X & Y \\ -Y & X \end{bmatrix} \begin{bmatrix} 1_n & -i 1_n \\ & 1_n \end{bmatrix})$$

$$= \det(\begin{bmatrix} X - iY & \\ -Y & X + iY \end{bmatrix}) = |\det(X + iY)|^2,$$

as required.

Exercise 1.6: If we know that $J(g, Z)$ is invertible, then (1.6) follows by definition of $g\langle Z \rangle$. Let us show non-singularity first for $Z = i 1_n$ and arbitrary g. If $J(g, Z)$ is singular, then considering $J(g, Z)^{\overline{t}} J(g, Z)$, we get that $C\,{}^t C + D\,{}^t D$ is also singular. But this matrix is symmetric and positive semi-definite. Hence, there is a nonzero column vector T such that ${}^t T (C\,{}^t C + D\,{}^t D) T = 0$. This implies that ${}^t T (C\,{}^t C) T = 0$ and ${}^t T (D\,{}^t D) T = 0$, which gives us ${}^t T C = {}^t T D = 0$. But this means that the rank of the matrix (C, D) is less than n, which is impossible, since g is non-singular. To get non-singularity for a general Z, first realize $Z = g\langle i 1_n \rangle$ for a suitable g and then use (1.6) for $Z = i 1_n$. This completes proof of part (*i*).

To get symmetry of $g\langle Z \rangle$, use (1.2) and (1.3) together with the relation

$$^t(CZ + D) g\langle Z \rangle (CZ + D) = Z\,{}^t C A Z + {}^t D A Z + Z\,{}^t C B + {}^t D B.$$

Let $Z' := g\langle Z \rangle$. Then

$$^t(C\bar{Z} + D)(Z' - \bar{Z}')(CZ + D)$$
$$= (\bar{Z}\,{}^t C + {}^t D)(AZ + B) - (\bar{Z}\,{}^t A + {}^t B)(CZ + D)$$
$$= \bar{Z}({}^t C A - {}^t A C)Z + ({}^t D A - {}^t B C)Z + \bar{Z}({}^t C B - {}^t A D)$$
$$= \mu(g)(Z - \bar{Z}).$$

This proves part (*ii*) of the theorem. Combining parts (*i*) and (*ii*) gives part (*iii*).

To get part (iv), we compute the Jacobian of the change of variable $Z \to g\langle Z \rangle$. For $Z = (z_{jk}) = (x_{jk} + iy_{jk}) \in \mathbb{H}_n$, set $Z' = (z'_{lm}) = (x'_{lm} + iy'_{lm}) = g\langle Z \rangle$. For $Z_1, Z_2 \in \mathbb{H}_n$, since Z'_2 is symmetric, we get

$$Z'_2 - Z'_1 = (Z_2 {}^tC + {}^tD)^{-1}(Z_2 {}^tA + {}^tB) - (AZ_1 + B)(CZ_1 + D)^{-1}$$
$$= \mu(g)(Z_2 {}^tC + {}^tD)^{-1}(Z_2 - Z_1)(CZ_1 + D)^{-1}$$

It follows that

$$\mathbf{D}Z' = \mu(g)(Z {}^tC + {}^tD)^{-1}\mathbf{D}Z(CZ + D)^{-1},$$

where $\mathbf{D}Z = (dz_{jk})$ and $\mathbf{D}Z' = (dz'_{lm})$. Note that if $\rho(U)$, with $U \in \mathrm{GL}_n(\mathbb{C})$, is the transformation $(v_{jk}) \mapsto U(v_{jk}) {}^tU$ of variables $v_{jk} = v_{kj}$ with $1 \leq j, k \leq n$, then $\det \rho(U) = \det(U)^{n+1}$. Let $\mathbf{d}Z$ and $\mathbf{d}Z'$ be the columns with entries $dz_{jk}(1 \leq j, k \leq n)$ and $dz'_{lm}(1 \leq l, m \leq n)$ arranged in a fixed order. Then, the above considerations imply the relation

$$\mathbf{d}Z' = \rho(\sqrt{\mu(g)} {}^t(CZ + D)^{-1})\mathbf{d}Z.$$

Taking $\mathbf{d}Z = \mathbf{d}X + i\mathbf{d}Y$, $\mathbf{d}Z' = \mathbf{d}X' + i\mathbf{d}Y'$, and $\rho(\sqrt{\mu(g)} {}^t(CZ + D)^{-1}) = R + iS$, we obtain

$$\mathbf{d}X' = R\mathbf{d}X - S\mathbf{d}Y, \qquad \mathbf{d}Y' = S\mathbf{d}X + R\mathbf{d}Y.$$

Thus, the Jacobian equals

$$\det\left(\begin{bmatrix} R & -S \\ S & R \end{bmatrix}\right) = \det\left(\begin{bmatrix} 1_n & i1_n \\ & 1_n \end{bmatrix}\begin{bmatrix} R & -S \\ S & R \end{bmatrix}\begin{bmatrix} 1_n & -i1_n \\ & 1_n \end{bmatrix}\right)$$
$$= \det\left(\begin{bmatrix} R + iS & \\ S & R - iS \end{bmatrix}\right) = \mu(g)^{n(n+1)}|\det(CZ + D)|^{-2n-2}.$$

This gives us part (iv) of the theorem. In addition, using part (ii), we get part (v) of the theorem.

Exercise 1.7: Take $g = \begin{bmatrix} & -1_n \\ 1_n & \end{bmatrix} \in \mathrm{GSp}_{2n}$. We know that $J(g, Z)$ is invertible from part (i) of Theorem 1.5. In this case, $J(g, Z) = Z$.

Exercise 1.9: All of the results are obtained by applying the automorphy conditions to F for various elements of $\gamma \in \Gamma_n$. For part (i) take $\gamma = -1_{2n}$. For part (ii) (a) take $\gamma = \begin{bmatrix} 1 & X \\ & 1 \end{bmatrix}$, for part (b) take $\gamma = \begin{bmatrix} g & \\ & {}^tg^{-1} \end{bmatrix}$, and for part (c) take $\gamma = \begin{bmatrix} & -1_n \\ 1_n & \end{bmatrix}$.

Exercise 1.10: Part (i) is clear from the hint. For part (ii), start with T which is not positive semi-definite. Hence, there is a primitive integral column vector x such that ${}^txTx < 0$. By Gauss lemma, there is a $g \in \mathrm{SL}_n(\mathbb{Z})$ whose first column is x. Hence, the $(1, 1)$ entry of the matrix ${}^tgTg$ is negative. This gives us part (ii). Set

$S_m := {}^t g_m T g_m$ with g_m as in the statement of the exercise. Choose a subsequence $\{m_v\}$ such that $m_v \to \infty$ as $v \to \infty$, and that the S_{m_v} are distinct. This is possible since the $(2, 2)$ entry of S_m is $t_{11}m^2 + t_{12}m + t_{22}$. Then

$$\mathrm{Tr}(S_{m_v}) = t_{11}m_v^2 + (\text{linear terms in } m_v) \to -\infty \text{ as } m_v \to \infty.$$

Hence, $e^{-2\pi \mathrm{Tr}(S_{m_v})} \to \infty$ as $m_v \to \infty$. This implies that $A(T) = 0$.

Chapter 2

Exercise 2.1: For $\gamma_0 = \begin{bmatrix} A' & B' \\ C' & D' \end{bmatrix} \in \Gamma_n$, we have

$$E_k^{(n)}(\gamma_0\langle Z \rangle) := \sum_{\left[\begin{smallmatrix} A & B \\ C & D \end{smallmatrix}\right] \in \Gamma_{0,n} \backslash \Gamma_n} \det(C\gamma_0\langle Z \rangle + D)^{-k}.$$

Let $\gamma = \begin{bmatrix} A & B \\ C & D \end{bmatrix}$. Then

$$\det(C\gamma_0\langle Z \rangle + D)^{-k} = \det(J(\gamma, \gamma_0\langle Z \rangle))^{-k} = \det(J(\gamma\gamma_0, Z)J(\gamma_0, Z)^{-1})^{-k}$$
$$= \det(C'Z + D')^k \det(J(\gamma\gamma_0, Z))^{-k}$$

Doing a change of variable $\gamma \to \gamma\gamma_0^{-1}$ in the summation for $E_k^{(n)}(Z)$, we get

$$E_k^{(n)}(\gamma_0\langle Z \rangle) = \det(C'Z + D')^k E_k^{(n)}(Z),$$

as required.

Exercise 2.2: Suppose $g = \begin{bmatrix} A' & 0 & B' & * \\ * & U & * & * \\ C' & 0 & D' & * \\ 0 & 0 & 0 & {}^t U^{-1} \end{bmatrix} \in P_r$ and $Z = \begin{bmatrix} Z_1 & Z' \\ {}^t Z' & Z_2 \end{bmatrix} \in \mathbb{H}_n$. Then

$$g\langle Z \rangle^* = \left(\left(\begin{bmatrix} A' & 0 \\ * & U \end{bmatrix} \begin{bmatrix} Z_1 & Z' \\ {}^t Z' & Z_2 \end{bmatrix} + \begin{bmatrix} B' & * \\ * & * \end{bmatrix} \right) \left(\begin{bmatrix} C' & 0 \\ 0 & 0 \end{bmatrix} \begin{bmatrix} Z_1 & Z' \\ {}^t Z' & Z_2 \end{bmatrix} + \begin{bmatrix} D' & * \\ 0 & {}^t U^{-1} \end{bmatrix} \right)^{-1} \right)^*$$
$$= (A'Z_1 + B')(C'Z_1 + D')^{-1}.$$

Hence, we have

$$f(g\langle Z \rangle^*) \det(CZ + D)^{-k} = \det(C'Z_1 + D')^k f(Z^*) \det(C'Z_1 + D')^{-k} = f(Z^*),$$

as required.

Exercise 2.4: We have

$$\dim(M_k(\Gamma_n)) = \dim(\mathrm{Ker}\,\Phi) + \dim(\mathrm{Im}\,\Phi) = \dim(S_k(\Gamma_n)) + \dim(M_k(\Gamma_{n-1})).$$

Now use that $\dim(S_k(\Gamma_n)) \le O(k^N)$, and induction on n to get the result.

Exercise 2.5: Since e_4 and e_6 generate the graded ring $M_*(\Gamma_1)$, and $\dim M_{10}(\Gamma_1) = 1$ we can take $M_{10}(\Gamma_1) = e_4 e_6 \mathbb{C}$. Since $e_{10} - e_4 e_6$ has weight 10, we must have $e_{10} - e_4 e_6 = \alpha e_4 e_6$. Comparing the constant term on both sides, we see that $\alpha = 0$. Hence, we get $e_{10} - e_4 e_6 = 0$. Applying the Siegel operator, we get $\Phi(\tilde{E}_{10}^{(2)} - \tilde{E}_4^{(2)} \tilde{E}_6^{(2)}) = e_{10} - e_4 e_6 = 0$. Hence, χ_{10} is a cusp form in $M_{10}(\Gamma_2)$.

Exercise 2.8: By definition of the Fourier coefficients of the Saito–Kurokawa lift, we have $A(\frac{nm}{d^2}, \frac{r}{d}, 1) = c(\det(2T)/d^2)$.

Exercise 2.9: Take $A = \mathrm{diag}(1, 1, 1, N)$.

Chapter 3

Exercise 3.1: Let $\gamma \in \Gamma_n$. Then

$$\bigsqcup_i \Gamma_n g_i \gamma = \Gamma_n g \Gamma_n \gamma = \Gamma_n g \Gamma_n = \bigsqcup_i \Gamma_n g_i.$$

Exercise 3.2: Note that $T_1 \cdot T_2$ is independent of the choice of $g \in \Gamma_n g$. Suppose we replace each g' by $\gamma' g'$ with $\gamma' \in \Gamma_n$. By definition of $\mathcal{H}_n$, we have for each γ',

$$\sum_g a_g \Gamma_n g \gamma' = T_1 \gamma' = T_1 = \sum_g a_g \Gamma_n g.$$

Hence,

$$\sum_{g,g'} a_g a_{g'} \Gamma_n g \gamma' g' = \sum_{g'} a_{g'} \left(\sum_g a_g \Gamma_n g \gamma'\right) g' = \sum_{g'} a_{g'} \left(\sum_g a_g \Gamma_n g\right) g' = \sum_{g,g'} a_g a_{g'} \Gamma_n g g'.$$

Finally, to see that $T_1 \cdot T_2$ is indeed an element of $\mathcal{H}_n$, let $\gamma \in \Gamma_n$.

$$(T_1 \cdot T_2)\gamma = \sum_{g,g'} a_g a_{g'} \Gamma_n g g' \gamma = T_1 \cdot (T_2 \gamma) = T_1 \cdot T_2,$$

as required. This gives us the well-definedness of the product.

Exercise 3.3: By the result on symplectic divisors, we can see that, for m, m' coprime, we have $T(mm') = T(m)T(m')$. Now, using the fundamental theorem of arithmetic, we can get the desired result.

Exercise 3.4: For part (i), we need two facts. The first one is that, for g_1, g_2, we have

$$\left(F|_k g_1\right)|_k g_2 = F|_k (g_1 g_2).$$

This can be checked by direct computation using the cocycle properties. Second, for every $\gamma \in \Gamma_n$, we have by definition of $\mathcal{H}_n$,

$$\Gamma_n g \Gamma_n = \sqcup_i \Gamma_n g_i = \sqcup_i \Gamma_n g_i \gamma.$$

Hence,

$$
\begin{aligned}
\left(T(g)F\right)(\gamma\langle Z\rangle) &= \sum_i (F|_k g_i)(\gamma\langle Z\rangle) = \sum_i \left((F|_k g_i)|_k \gamma\right)(Z) \det(J(\gamma, Z))^k \\
&= \sum_i (F|_k g_i \gamma)(Z) \det(J(\gamma, Z))^k = \det(J(\gamma, Z))^k \sum_i (F|_k g_i)(Z).
\end{aligned}
$$

For part (ii), let $\Gamma_n g \Gamma_n = \sqcup_i \Gamma_n g_i$. Set $\mu = \mu(g)$. Taking inverses and multiplying by μ, we get $\Gamma_n \mu g^{-1} \Gamma_n = \sqcup_i \mu g_i^{-1} \Gamma_n$. By the result on symplectic divisors, we can take g to be a diagonal matrix. Then, we can see that $\mu g^{-1} = J g J$. Hence,

$$\sqcup_i \mu g_i^{-1} \Gamma_n = \Gamma_n \mu g^{-1} \Gamma_n = \Gamma_n g \Gamma_n = \sqcup_i \Gamma_n g_i.$$

We have

$$
\begin{aligned}
\langle T(g)F, G\rangle &= \sum_i \langle F|_k g_i, G\rangle = \sum_i \langle F|_k g_i, G|_k g_i^{-1}|_k g_i\rangle \\
&= \sum_i \mu(g_i)^{n(k-n-1)} \langle F, G|_k g_i^{-1}\rangle = \langle F, \sum_i \mu(g_i)^{n(k-n-1)} G|_k g_i^{-1}\rangle \\
&= \langle F, \sum_i G|_k(\mu g_i^{-1})\rangle = \langle F, \sum_i G|_k g_i\rangle = \langle F, T(g)G\rangle.
\end{aligned}
$$

To make sense of the inner product of functions that are not invariant under Γ_n, we extend the Petersson inner product to

$$\langle F', G'\rangle = \frac{1}{[\Gamma_n : \Gamma_n(N)]} \int_{\Gamma_n(N)\backslash \mathbb{H}_n} F'(Z)\overline{G'(Z)} \det(Y)^k \frac{dX\, dY}{\det(Y)^{n+1}}$$

where F', G' are both modular forms with respect to the congruence subgroup $\Gamma_n(N)$ (defined in (2.13)) for some integer N. In addition, we have used $\mu(g_i) = \mu$ for all i and the $\mathrm{GSp}_{2n}(\mathbb{R})^+$-invariance of the measure on $\mathbb{H}_n$.

Exercise 3.7: Let us apply the slash operator to the various single coset representatives one at a time, and calculate the contribution toward the Fourier coefficient. First consider

$$\left(F|_k \begin{bmatrix} p1_2 & \\ & 1_2 \end{bmatrix}\right)(Z) = p^{2k-3} F(pZ) = p^{2k-3} \sum_{\substack{{}^tT=T>0 \\ \text{half-integral}}} A(T)e^{2\pi i \operatorname{Tr}(TpZ)}$$

$$= p^{2k-3} \sum_{\substack{{}^tT=T>0 \\ \text{half integral} \\ p|T}} A(\tfrac{1}{p}T)e^{2\pi i \operatorname{Tr}(TZ)}.$$

Hence, the contribution to the Fourier coefficient is $p^{2k-3}A(1/pT)$. Next consider

$$\sum_{a,b,d\in\mathbb{Z}/p\mathbb{Z}} \left(F|_k \begin{bmatrix} 1 & a & b \\ & 1 & b & d \\ & & p & \\ & & & p \end{bmatrix}\right)(Z)$$

$$= p^{2k-3}p^{-2k} \sum_{a,b,d\in\mathbb{Z}/p\mathbb{Z}} F(\tfrac{1}{p}Z + \tfrac{1}{p}\begin{bmatrix} a & b \\ b & d \end{bmatrix})$$

$$= p^{-3} \sum_{a,b,d\in\mathbb{Z}/p\mathbb{Z}} \left(\sum_{\substack{{}^tT=T>0 \\ \text{half integral}}} A(T)e^{2\pi i \operatorname{Tr}(T\frac{1}{p}Z)} e^{2\pi i \operatorname{Tr}(T\frac{1}{p}\begin{bmatrix} a & b \\ c & d \end{bmatrix})} \right).$$

If $T = \begin{bmatrix} n & r/2 \\ r/2 & m \end{bmatrix}$, then $\operatorname{Tr}(T\begin{bmatrix} a & b \\ b & d \end{bmatrix}) = an + rb + md$. It is easy to check that

$$\sum_{a,b,d\in\mathbb{Z}/p\mathbb{Z}} e^{2\pi i \frac{1}{p}(an+rb+md)} = \begin{cases} p^3 & \text{if } p|n,\ p|r,\ p|m; \\ 0 & \text{otherwise.} \end{cases}$$

Hence, the contribution toward the Fourier coefficient is $A(pT)$. Next we have

$$\sum_{a\in\mathbb{Z}/p\mathbb{Z}} \left(F|_k \begin{bmatrix} 1 & & a & \\ & p & & \\ & & p & \\ & & & 1 \end{bmatrix}\right)(Z)$$

$$= p^{2k-3}p^{-k} \sum_{a\in\mathbb{Z}/p\mathbb{Z}} F\left(\left(\begin{bmatrix} 1 & \\ & p \end{bmatrix}Z + \begin{bmatrix} a & \\ & 1 \end{bmatrix}\right)\begin{bmatrix} p & \\ & 1 \end{bmatrix}^{-1}\right)$$

$$= p^{k-3} \sum_{a\in\mathbb{Z}/p\mathbb{Z}} F\left(\begin{bmatrix} 1 & \\ & p \end{bmatrix}Z\begin{bmatrix} p & \\ & 1 \end{bmatrix}^{-1} + \begin{bmatrix} a/p & \\ & 1 \end{bmatrix}\right)$$

$$= p^{k-3} \sum_{a\in\mathbb{Z}/p\mathbb{Z}} \sum_{\substack{{}^tT=T>0 \\ \text{half-integral}}} A(T)e^{2\pi i \operatorname{Tr}(T\begin{bmatrix} 1 & \\ & p \end{bmatrix}Z\begin{bmatrix} p^{-1} & \\ & 1 \end{bmatrix})} e^{2\pi i \operatorname{Tr}(T\begin{bmatrix} a/p & \\ & 1 \end{bmatrix})}$$

Once again, if $T = \begin{bmatrix} n & r/2 \\ r/2 & m \end{bmatrix}$, then $\mathrm{Tr}(T\begin{bmatrix} a/p & \\ & 1 \end{bmatrix}) = na/p + m$. Hence,

$$\sum_{a \in \mathbb{Z}/p\mathbb{Z}} e^{2\pi i \mathrm{Tr}(T\begin{bmatrix} a/p & \\ & 1 \end{bmatrix})} = \begin{cases} p & \text{if } p|n; \\ 0 & \text{otherwise.} \end{cases}$$

Note that $p|n$ if and only if $\begin{bmatrix} p^{-1} & \\ & 1 \end{bmatrix} T \begin{bmatrix} 1 & \\ & p \end{bmatrix}$ is half-integral. So, we get

$$\sum_{a \in \mathbb{Z}/p\mathbb{Z}} (F|_k \begin{bmatrix} 1 & a \\ & p & \\ & & p \\ & & & 1 \end{bmatrix})(Z)$$

$$= p^{k-2} \sum_{\substack{{}^t T = T > 0 \\ T \text{ half-integral} \\ \begin{bmatrix} p^{-1} & \\ & 1 \end{bmatrix} T \begin{bmatrix} 1 & \\ & p \end{bmatrix} \text{half-integral}}} A(T) e^{2\pi i \mathrm{Tr}(\begin{bmatrix} p^{-1} & \\ & 1 \end{bmatrix} T \begin{bmatrix} 1 & \\ & p \end{bmatrix} Z)}$$

$$= p^{k-2} \sum_{\substack{{}^t T = T > 0 \\ \text{half-integral}}} A(\begin{bmatrix} p & \\ & 1 \end{bmatrix} T \begin{bmatrix} 1 & \\ & p^{-1} \end{bmatrix}) e^{2\pi i \mathrm{Tr}(TZ)}.$$

Hence, the contribution toward the Fourier coefficient is $p^{k-2} A(\frac{1}{p} \begin{bmatrix} p & \\ & 1 \end{bmatrix} T \begin{bmatrix} p & \\ & 1 \end{bmatrix})$. The last computation is similar as well, and we get the final contribution to the Fourier coefficient of $p^{k-2} \sum_{a \in \mathbb{Z}/p\mathbb{Z}} A(\frac{1}{p} \begin{bmatrix} 1 & \alpha \\ & p \end{bmatrix} T \begin{bmatrix} 1 & \\ \alpha & p \end{bmatrix})$.

Exercise 3.9: We know that

$$Q(X) = (1 - \alpha_0 X)(1 - \alpha_0 \alpha_1 X)(1 - \alpha_0 \alpha_2 X)(1 - \alpha_0 \alpha_1 \alpha_2 X).$$

Comparing coefficient of X^4, we only get $\alpha_0^2 \alpha_1 \alpha_2 = \pm p^{2k-3}$. To get the plus sign, compare coefficients of X and X^3.

Exercise 3.10: Follows from Theorem 3.6.

Exercise 3.12: We have

$$Q(X) = 1 - \lambda(p)X + (\lambda(p)^2 - \lambda(p^2) - p^{2k-4})X^2 - \lambda(p)p^{2k-3}X^3 + p^{4k-6}X^4$$
$$= (1 - \alpha_{0,p}X)(1 - \alpha_{0,p}\alpha_{1,p}X)(1 - \alpha_{0,p}\alpha_{2,p}X)(1 - \alpha_{0,p}\alpha_{1,p}\alpha_{2,p}X).$$

Now comparing the coefficients of X and X^2, we get the result.

Exercise 3.13: Recall the doubling formula for the gamma function

$$\Gamma(z)\Gamma(z + \frac{1}{2}) = 2^{1-2z}\sqrt{\pi}\Gamma(2z),$$

and the formula $\Gamma(z + 1) = z\Gamma(z)$. Using these, we get

$$\Lambda(s, F_f) = \pi^{1-k}2^{-k}(s - k + 1)\Lambda(s, f)\xi(s - k + 1)\xi(s - k + 2).$$

Now, using the functional equations for ξ and $L(s, f)$ and the assumption that k is even, we get

$$(-1)^k\Lambda(2k - 2 - s, F_f)$$
$$= \pi^{1-k}2^{-k}(2k - 2 - s - k + 1)\Lambda(2k - 2 - s, f)$$
$$\times \xi(2k - 2 - s - k + 1)\xi(2k - 2 - s - k + 2)$$
$$= \pi^{1-k}2^{-k}(-s + k - 1)(-1)^{k-1}\Lambda(s, f)$$
$$\times \xi(1 - (2k - 2 - s - k + 1))\xi(1 - (2k - 2 - s - k + 2))$$
$$= \pi^{1-k}2^{-k}(s - k + 1)\Lambda(s, f)\xi(s - k + 1)\xi(s - k + 2)$$
$$= \Lambda(s, F_f).$$

Exercise 3.14: From the relation

$$L(s, F_f, \text{spin}) = \zeta(s - k + 1)\zeta(s - k + 2)L(s, f),$$

we have

$$\{\alpha_{0,p}, \alpha_{0,p}\alpha_{1,p}, \alpha_{0,p}\alpha_{2,p}, \alpha_{0,p}\alpha_{1,p}\alpha_{2,p}\} = \{p^{k-1}, p^{k-2}, \beta_{0,p}, \beta_{0,p}\beta_{1,p}\}.$$

Here, $\beta_{0,p}, \beta_{1,p}$ are local Satake p-parameters of f and we have $|\beta_{0,p}| = p^{k-3/2}$. Using $\alpha_{0,p}^2\alpha_{1,p}\alpha_{2,p} = p^{2k-3}$, we deduce that $\{\alpha_{1,p}, \alpha_{2,p}\} = \{p^{-(k-1)}\beta_{0,p}, p^{-(k-2)}\beta_{0,p}\}$. Note that $\beta_{0,p}^2\beta_{1,p} = p^{2k-3}$ implies that $\left(p^{-(k-1)}\beta_{0,p}\right)^{-1} = p^{-(k-2)}\beta_{0,p}\beta_{1,p}$ and $\left(p^{-(k-2)}\beta_{0,p}\right)^{-1} = p^{-(k-1)}\beta_{0,p}\beta_{1,p}$. Putting this into the definition of the standard L-function, we get the result.

Chapter 4

Exercise 4.4: Follow from Theorems 3.8 and 4.3.

Exercise 4.6:

(i) Let $\gamma_1 = \begin{bmatrix} a & b \\ & 1 & \\ c & d \\ & & 1 \end{bmatrix}$ and $\gamma_2 = \begin{bmatrix} 1 & \\ s & 1 & \\ & 1 & -s \\ & & 1 \end{bmatrix}$. Let $Z = \begin{bmatrix} \tau & z \\ z & \tau' \end{bmatrix}$. Then

$$\gamma_1 \langle Z \rangle = (\begin{bmatrix} a & \\ & 1 \end{bmatrix} \begin{bmatrix} \tau & z \\ z & \tau' \end{bmatrix} + \begin{bmatrix} b & 0 \\ 0 & 0 \end{bmatrix})(\begin{bmatrix} c & 0 \\ 0 & 0 \end{bmatrix} \begin{bmatrix} \tau & z \\ z & \tau' \end{bmatrix} + \begin{bmatrix} d & \\ & 1 \end{bmatrix})^{-1}$$

$$= \begin{bmatrix} a\tau + b & az \\ z & \tau' \end{bmatrix} \begin{bmatrix} c\tau + d & cz \\ & 1 \end{bmatrix}^{-1} = \begin{bmatrix} a\tau + b & az \\ z & \tau' \end{bmatrix} \begin{bmatrix} (c\tau + d)^{-1} & -cz(c\tau + d)^{-1} \\ & 1 \end{bmatrix}$$

$$= \begin{bmatrix} \frac{a\tau+b}{c\tau+d} & \frac{z}{c\tau+d} \\ \frac{z}{c\tau+d} & \tau' - \frac{cz^2}{c\tau+d} \end{bmatrix}$$

We also have

$$\gamma_2 \langle Z \rangle = \begin{bmatrix} 1 & \\ s & 1 \end{bmatrix} \begin{bmatrix} \tau & z \\ z & \tau' \end{bmatrix} \begin{bmatrix} 1 & s \\ & 1 \end{bmatrix} = \begin{bmatrix} \tau & z + s\tau \\ z + s\tau & \tau' + 2sz + s^2\tau \end{bmatrix}.$$

Also, $\det J(\gamma_1, Z) = c\tau + d$ and $\det J(\gamma_2, Z) = 1$.

(ii) Writing the Fourier–Jacobi expansion of both the sides of the equation in part (i), we get

$$\sum_{m=1}^{\infty} \phi_m(\frac{a\tau + b}{c\tau + d}, \frac{z}{c\tau + d}) e^{2\pi i m\tau'} e^{\frac{2\pi i m c z^2}{c\tau+d}} = (c\tau + d)^k \sum_{m=1}^{\infty} \phi_m(\tau, z) e^{2\pi i m\tau'},$$

and

$$\sum_{m=1}^{\infty} \phi_m(\tau, z + s\tau) e^{2\pi i m\tau'} e^{2\pi i m(2sz + s^2\tau)} = \sum_{m=1}^{\infty} \phi_m(\tau, z) e^{2\pi i m\tau'}.$$

Comparing the coefficients of $e^{2\pi i m\tau'}$ on both sides, we get the problem.

(iii) We have

$$\phi_m(\tau, z) = \sum_{\substack{n, r \in \mathbb{Z}, n > 0 \\ 4mn - r^2 > 0}} A(n, r, m) e^{2\pi i n\tau} e^{2\pi i r z}.$$

The expression

$$\frac{\partial^\nu}{\partial z^\nu} e^{2\pi i r z}|_{z=0} = (2\pi i r)^\nu$$

implies

$$\lambda_\nu(\tau, z) = \frac{1}{\nu!} \sum_{\substack{n,r\in\mathbb{Z}, n>0 \\ 4mn-r^2>0}} (2\pi i r)^\nu A(n, r, m) e^{2\pi i n\tau} z^\nu.$$

(iv) We have

$$\sum_{\nu=\nu_0}^{\infty} \frac{\lambda_\nu(\frac{a\tau+b}{c\tau+d}, z)}{(c\tau + d)^{k+\nu}} = \frac{1}{(c\tau + d)^k} \sum_{\nu=\nu_0}^{\infty} \lambda_\nu(\frac{a\tau + b}{c\tau + d}, \frac{z}{c\tau + d})$$

$$= \frac{1}{(c\tau + d)^k} \phi_m(\frac{a\tau + b}{c\tau + d}, \frac{z}{c\tau + d})$$

$$= e^{\frac{2\pi i m c z^2}{c\tau+d}} \phi_m(\tau, z)$$

$$= \Big(\sum_{j=0}^{\infty} \frac{1}{j!}\big(\frac{2\pi i m c z^2}{c\tau + d}\big)^j\Big) \sum_{\nu=\nu_0}^{\infty} \lambda_\nu(\tau, z)$$

Now, comparing the coefficients of z^{ν_0} on both sides, we get

$$\lambda_{\nu_0}(\frac{a\tau + b}{c\tau + d}, z) = (c\tau + d)^{k+\nu_0} \lambda_{\nu_0}(\tau, z),$$

as required.

(v) Assume that $A(n, r, m) = 0$ whenever $\gcd(n, r, m) = 1$. Note that this implies $\phi_1 = 0$. Let $m > 1$ be such that $\phi_m \neq 0$. Hence, the corresponding $\lambda_{\nu_0} \neq 0$. For any fixed z, we know that $\lambda_{\nu_0}(\tau, z) \in M_{k+\nu_0}(\Gamma_1)$ and we can see that the nth Fourier coefficient is given by

$$\frac{1}{\nu_0!} \sum_{\substack{r\in\mathbb{Z} \\ 4mn-r^2>0}} (2\pi i r)^{\nu_0} A(n, r, m) z^{\nu_0}$$

The hypothesis on $A(n, r, m)$ tells us that the above nth Fourier coefficient vanishes for all n coprime to m. From the given fact of modular forms, we can conclude that $\lambda_{\nu_0}(\tau, z) = 0$, a contradiction.

Exercise 4.10: If $\{c(n)\}$ are the Fourier coefficients of the half-integral weight form g, then the plus space condition implies that $c(n) = 0$ if $n \equiv 1, 2 \pmod 4$. Here, we have used that k is even. Now, let N be a positive integer such that $c(N) \neq 0$. We have

$$N = \det(2T), \text{ with } T = \begin{cases} \begin{bmatrix} a+1 & 1/2 \\ 1/2 & 1 \end{bmatrix} & \text{if } N = 4a + 3; \\ \begin{bmatrix} a & \\ & 1 \end{bmatrix} & \text{if } N = 4a. \end{cases}$$

In both the above cases, the definition of the Fourier coefficient of the Saito–Kurokawa lift gives $A(T) = c(N) \neq 0$.

Chapter 5

Exercise 5.2: Let $T(g) = \Gamma_n g \Gamma_n = \bigsqcup_i \Gamma_n g_i$ and choose representatives g_i of the form $\begin{bmatrix} A_i & B_i \\ & D_i \end{bmatrix}$. Let the Fourier coefficients of $T(g)F$ be given by $\{B(T)\}$. Then, one can show that

$$B(T) = \sum_i \det(D_i)^{-k} A(D_i T A_i^{-1}) e^{2\pi i \operatorname{Tr}(T A_i^{-1} B_i)}.$$

Using $|A(T)| \ll_F \det(T)^{k/2}$, and the fact that, for any g there are only finitely many choices of A_i, B_i, D_i, we can conclude that $|B(T)| \leq C \det(T)^{k/2}$ where the constant C depends on F and g but not T. This shows that the action of the Hecke algebra stabilizes the modular forms satisfying $|A(T)| \ll_F \det(T)^{k/2}$. Consider the subspace V_F of $M_k(\Gamma_n)$ spanned by $\{T(g)F : g \in G_n\}$. We can find a basis of Hecke eigenforms for V_F, and any one of these basis elements gives a solution to the first part of the problem. If F is not a cusp form, then not every element of the basis can be a cusp form. Such a non-cuspidal basis element gives the second part of the problem.

Exercise 5.3: Suppose $\Phi^n F = 0$. By induction, we can see that

$$L(s, F, \text{std}) = \prod_{i=0}^{n-1} \zeta(s - k + n - i)\zeta(s + k - n + i).$$

The rightmost pole is at $s = k$, and it is not canceled by the zeros of the other terms. Since $k > 2n$, we get a contradiction to the holomorphy of $L(s, F, \text{std})$.

Exercise 5.4: We have $\langle T(g)F, G \rangle = \langle F, T(g)G \rangle$, which shows that the Hecke eigenvalues are real.

Exercise 5.6: We have $L(s, F_f, \text{spin}) = \zeta(s - k + 1)\zeta(s - k + 2)L(s, f)$. Write $L(s, f) = \prod_p (1 - \omega(p)p^{-s} + p^{2k-3-2s})^{-1}$. Write

$$L(s, F_f, \text{spin}) = \sum_{n=1}^{\infty} \frac{a(n)}{n^s} = \prod_p \sum_{n=0}^{\infty} \frac{a(p^n)}{p^{ns}}.$$

For a fixed prime p, we have the formal identity in an indeterminate X

$$\sum_{n=0}^{\infty} a(p^n)X^n = \frac{1}{(1 - p^{k-1}X)(1 - p^{k-2}X)(1 - \omega(p)X + p^{2k-3}X^2)}.$$

The Ramanujan estimate for genus 1 gives us the factorization $1 - \omega(p)X + p^{2k-3}X^2 = (1 - \alpha p^{k-3/2}X)(1 - \bar{\alpha}p^{k-3/2}X)$ for a complex number α satisfying $|\alpha| = 1$. Let us assume that $\alpha \neq \bar{\alpha}$ in the following. The $\alpha = \pm 1$ case needs appropriate modifications.

$$\sum_{n=0}^{\infty} \frac{a(p^n)}{(p^{k-3/2})^n}X^n = \frac{1}{(1 - \sqrt{p}X)(1 - \frac{1}{\sqrt{p}}X)(1 - \alpha X)(1 - \bar{\alpha}X)}$$

$$= \frac{1}{|\sqrt{p} - \alpha|^2}\left(\frac{p^2}{(p-1)(1 - \sqrt{p}X)} - \frac{1}{(p-1)(1 - \frac{1}{\sqrt{p}}X)}\right.$$

$$+ \sqrt{p}\left(\frac{1}{\bar{\alpha}(\bar{\alpha}^2 - 1)(1 - \alpha X)} + \frac{1}{\alpha(\alpha^2 - 1)(1 - \bar{\alpha}X)}\right)\right)$$

Expand as geometric series and compare coefficients of X^n to get

$$\frac{a(p^n)}{(p^{k-3/2})^n}|\sqrt{p} - \alpha|^2 = \frac{p^2}{p-1}\sqrt{p^n} - \frac{1}{p-1}\frac{1}{\sqrt{p^n}} - \sqrt{p}\left(\frac{\alpha^n}{\bar{\alpha}(1 - \bar{\alpha}^2)} + \frac{\bar{\alpha}^n}{\alpha(1 - \alpha^2)}\right)$$

$$= \frac{p^2}{p-1}\sqrt{p^n} - \frac{1}{p-1}\frac{1}{\sqrt{p^n}} - \sqrt{p}(\alpha^{n+1} + \alpha^{n-1} + \cdots + \frac{1}{\alpha^{n+1}})$$

Using $|\alpha| = 1$, we get the estimate for all $n \in \mathbb{N}$

$$\frac{a(p^n)}{(p^{k-3/2})^n}|\sqrt{p} - \alpha|^2 \geq \frac{p^2}{p-1}\sqrt{p^n} - \frac{1}{p-1}\frac{1}{\sqrt{p^n}} - \sqrt{p}(n + 2).$$

This gives us $a(p^n) > 0$ for all n. To get to $\lambda(n)$, use $L(s, F, \text{spin}) = \zeta(2s - 2k + 4)\sum \lambda(n)n^{-s}$, to get

$$\frac{\lambda(p^n)}{(p^{k-3/2})^n} = \frac{a(p^n)}{(p^{k-3/2})^n} - \frac{1}{p}\frac{a(p^{n-2})}{(p^{k-3/2})^{n-2}}$$

for $n \geq 2$ and $\lambda(p^n) = a(p^n)$ for $n = 0, 1$. Hence, we have

$$\frac{\lambda(p^n)}{(p^{k-3/2})^n}|\sqrt{p} - \alpha|^2 = (p+1)\sqrt{p^n} - \sqrt{p}((1 - \frac{1}{p})(\alpha^{n-1} + \cdots + \frac{1}{\alpha^{n-1}}) + \alpha^{n+1} + \frac{1}{\alpha^{n+1}})$$

$$\geq (p+1)\sqrt{p^n} - \sqrt{p}(n + 2).$$

This immediately implies that $\lambda(p^n) > 0$ for all $n \geq 0$, and hence, $\lambda(n) > 0$ for all $n \in \mathbb{N}$.

Exercise 5.9: Substituting the Fourier expansion of F, we get

$$\int_{R_2} F(iY) \det Y^{s-3/2} dY = \int_{R_2} \det Y^{s-3/2} \sum_{T>0} A(T) e^{-2\pi \mathrm{Tr}(TY)} dY$$

$$= \sum_{\{T\}>0} \frac{A(T)}{\epsilon(T)} \sum_{g \in GL_2(\mathbb{Z})} \int_{R_2} \det Y^{s-3/2} e^{-2\pi \mathrm{Tr}({}^t gTgY)} dY$$

$$= \sum_{\{T\}>0} \frac{A(T)}{\epsilon(T)} \int_{P_2} \det Y^{s-3/2} e^{-2\pi \mathrm{Tr}(TY)} dY.$$

This completes part (1) of the problem. Next, note that

$$\det(Y) = y_1 y_2, \quad dY = y_1 dy_1 dy_2 dy_3, \quad \mathrm{Tr}(TY) = t_1 y_1 + (y_2 + y_1 y_3^2) t_2.$$

This implies

$$\int_{P_2} \det Y^{s-3/2} e^{-2\pi \mathrm{Tr}(TY)} dY$$

$$= \int_{y_1>0, y_2>0, y_3} y_1^{s-1/2} e^{-2\pi t_1 y_1} y_2^{s-3/2} e^{-2\pi y_2 t_2} e^{-2\pi y_3^2 y_1 t_2} dy_1 dy_2 dy_3,$$

which gives us part (2) of the problem. Using the integral formulas in part (3), the integral on the right-hand side reduces to

$$\frac{\Gamma(s-\frac{1}{2})}{(2\pi t_2)^{(s-\frac{1}{2})}} \sqrt{\frac{2}{t_2}} \int_0^\infty y_1^{s-1} e^{-2\pi t_1 y_1} dy_1 = \frac{\Gamma(s-\frac{1}{2})}{(2\pi t_2)^{(s-\frac{1}{2})}} \sqrt{\frac{2}{t_2}} \frac{\Gamma(s)}{(2\pi t_1)^s}$$

$$= 2(2\pi)^{-2s} \pi^{\frac{1}{2}} \frac{\Gamma(s)\Gamma(s-\frac{1}{2})}{\det(T)^s}.$$

Substituting this in part (1), we get the result of part (3).

Chapter 6

Exercise 6.1: The strong approximation for Sp_{2n} states that

$$\mathrm{Sp}_{2n}(\mathbb{A}) = \mathrm{Sp}_{2n}(\mathbb{Q})\mathrm{Sp}_{2n}(\mathbb{R}) \prod_{p<\infty} \mathrm{Sp}_{2n}(\mathbb{Z}_p).$$

Given any $g \in G(\mathbb{A})$, we can write it as

$$g = \begin{bmatrix} 1_n & \\ & \mu(g)1_n \end{bmatrix} g',$$

where $g' \in \mathrm{Sp}_{2n}(\mathbb{A})$, and $\mu(g) \in \mathbb{A}^\times$ is the similitude of g. Now, the exercise follows from the decomposition of $\mathbb{A}^\times$ given by

$$\mathbb{A}^\times = \mathbb{Q}^\times \times \mathbb{R}_+^\times \times \prod_{p<\infty} \mathbb{Z}_p^\times.$$

Exercise 6.2: Write $k_0' = \otimes_{p<\infty} k_{0,p}$, with $k_{0,p} \in G(\mathbb{Z}_p)$. We have

$$g_\infty \otimes_{p<\infty} 1 = g_\mathbb{Q}' g_\infty' k_0' = \left(g_\mathbb{Q}' \otimes_{p<\infty} g_\mathbb{Q}'\right)\left(g_\infty' \otimes_{p<\infty} k_{0,p}\right) = g_\mathbb{Q}' g_\infty' \otimes_{p<\infty} g_\mathbb{Q}' k_{0,p}$$

Comparing the archimedean components, we get $g_\infty = g_\mathbb{Q}' g_\infty'$. Comparing the p-adic components, we get $g_\mathbb{Q}' \in G(\mathbb{Z}_p)$ for all $p < \infty$. Hence, $g_\mathbb{Q}' \in G(\mathbb{Z}) = \Gamma_n$. This gives us part (i). To show well-definedness, suppose $g = \gamma g_\infty k = \gamma' g_\infty' k'$. Then, from part (i), we have $g_\infty = (\gamma^{-1}\gamma')g_\infty'$ and $\gamma^{-1}\gamma' \in \Gamma_n$. Hence,

$$(F||_k g_\infty)(I) = (F||_k(\gamma^{-1}\gamma')g_\infty')(I)$$
$$= ((F||_k(\gamma^{-1}\gamma')||_k g_\infty')(I)$$
$$= (F||_k g_\infty')(I).$$

Here, we have used that $F||_k \gamma = F$ for all $\gamma \in \Gamma_n$. This shows us that Φ_F is well defined.

Exercise 6.5: Observe that, since S is assumed to be half-integral, $\theta_S(n) = 1$, for all $n \in \prod_{p<\infty} U(\mathbb{Z}_p)$. Using the strong approximation of the ring of adeles $\mathbb{A}$, we see that $U(\mathbb{A}) = U(\mathbb{Q}) \times U(\mathbb{R}) \times \prod_{p<\infty} U(\mathbb{Z}_p)$. Hence, we get the following fundamental domain:

$$U(\mathbb{Q})\backslash U(\mathbb{A}) \simeq \left(U(\mathbb{Z})\backslash U(\mathbb{R})\right) \times \prod_{p<\infty} U(\mathbb{Z}_p).$$

Recall Φ_F is right invariant under $\prod_{p<\infty} U(\mathbb{Z}_p)$ as well. Denote by Sym_2 the set of symmetric matrices and P_2 as the set of positive definite symmetric half-integral matrices. We get

$$\Phi_F^S(1) = \int\limits_{U(\mathbb{Z})\backslash U(\mathbb{R})} \Phi_F(n)\theta_S^{-1}(n)\,dn$$

$$= \int\limits_{Sym_2(\mathbb{Z})\backslash Sym_2(\mathbb{R})} (F\|_k\begin{bmatrix} 1 & X \\ & 1 \end{bmatrix})(I)\theta_S^{-1}(\begin{bmatrix} 1 & X \\ & 1 \end{bmatrix})dX$$

$$= \int\limits_{Sym_2(\mathbb{Z})\backslash Sym_2(\mathbb{R})} F(I+X)e^{-2\pi i \mathrm{Tr}(SX)}dX$$

$$= \sum_{T\in P_2} A(T)e^{2\pi i \mathrm{Tr}(TI)} \int\limits_{Sym_2(\mathbb{Z})\backslash Sym_2(\mathbb{R})} e^{2\pi i \mathrm{Tr}(TX)}e^{-2\pi i \mathrm{Tr}(SX)}dX$$

$$= A(S)e^{-2\pi\,\mathrm{Tr}(S)}.$$

Exercise 6.6: We have

$$\Phi_{F_1}(g)\overline{\Phi_{F_2}(g)} = F_1(g\langle I\rangle)\overline{F_2(g\langle I\rangle)}\mu(g)^{nk}|\det J(g,I)|^{-2k}$$
$$= F_1(Z)\overline{F}_2(Z)\det Y^k$$

where $g \in G_\infty^+$ and $g\langle I\rangle = Z = X + iY$. Define $f(Z) = F_1(Z)\overline{F}_2(Z)\det Y^k$. Then f is Γ_n-invariant and $\Phi_f(g) = \Phi_{F_1}(g)\overline{\Phi_{F_2}(g)}$. Using the relation between Haar measures, we get

$$\langle F_1, F_2\rangle = \int\limits_{\Gamma_n\backslash \mathbb{H}_n} f(Z)d^*Z = \int\limits_{Z(\mathbb{A})G(\mathbb{Q})\backslash G(\mathbb{A})} \Phi_f(g)dg = \langle\Phi_{F_1}, \Phi_{F_2}\rangle.$$

Exercise 6.7: Let $f \in \mathcal{H}(G_p, K_p)$. By definition of $\mathcal{H}(G_p, K_p)$, f is invariant on double cosets $K_p g K_p$ for $g \in G_p$. Since f has compact support, there are finitely many g_i such that $\mathrm{supp}(f) = \sqcup_i K_p g_i K_p$. If $f(g_i) = a_i \in \mathbb{C}$, then we have

$$f(g) = \sum_i a_i \mathrm{char}(K_p g_i K_p)(g).$$

Exercise 6.10: Note that

$$(X_1 * X_1)(x) = \int\limits_{T_p} X_1(xt)X_1(t^{-1})dt = \int\limits_{\mathrm{diag}(p^{-1}\mathbb{Z}_p^\times,\mathbb{Z}_p^\times,\cdots,\mathbb{Z}_p^\times,p\mathbb{Z}_p^\times,\mathbb{Z}_p^\times,\cdots,\mathbb{Z}_p^\times)} X_1(xt)dt$$

$$= \begin{cases} 1 & \text{if } x \in \mathrm{diag}(p^2\mathbb{Z}_p^\times,\mathbb{Z}_p^\times,\cdots,\mathbb{Z}_p^\times,p^{-2}\mathbb{Z}_p^\times,\mathbb{Z}_p^\times,\cdots,\mathbb{Z}_p^\times) \\ 0 & \text{otherwise.} \end{cases}$$

$$= \mathrm{char}(\mathrm{diag}(p^2\mathbb{Z}_p^\times,\mathbb{Z}_p^\times,\cdots,\mathbb{Z}_p^\times,p^{-2}\mathbb{Z}_p^\times,\mathbb{Z}_p^\times,\cdots,\mathbb{Z}_p^\times))(x).$$

Here, we have assumed that the measure is normalized so that $\mathbb{Z}_p^\times$ has volume 1. Similar calculation will show that

$$\mathrm{char}(\mathrm{diag}(p^{r_1}\mathbb{Z}_p^\times, p^{r_2}\mathbb{Z}_p^\times, \cdots, p^{r_n}\mathbb{Z}_p^\times, p^{r-r_1}\mathbb{Z}_p^\times, p^{r-r_2}\mathbb{Z}_p^\times, \cdots, p^{r-r_n}\mathbb{Z}_p^\times))$$
$$= X_0^r X_1^{r_1} \cdots X_n^{r_n}.$$

This gives us the result.

Exercise 6.12: Let $f \in \mathrm{Ind}_B^{G_p}(\chi)$ and let $g_0 = z1_{2n} \in Z(\mathbb{Q}_p)$. Note that $g_0 \in T(\mathbb{Q}_p)$. Then

$$(\pi(g_0)f)(g) = f(gg_0) = f(g_0g) = |\delta_B(g_0)|^{1/2}\chi(g_0)f(g).$$

We have $g_0 = \mathrm{diag}(u_1, u_2, \cdots, u_n, u_1^{-1}u_0, u_2^{-1}u_0, \cdots, u_n^{-1}u_0)$, with $u_0 = z^2$ and $u_i = z$ for $i = 1, \cdots, n$. Hence

$$|\delta_B(g_0)|^{1/2}\chi(g_0) = |z^{-n(n+1)/2}zz^2 \cdots z^n|\chi_0(z^2)\chi_1(z) \cdots \chi_n(z) = (\chi_0^2\chi_1 \cdots \chi_n)(z).$$

Hence, the center acts on the representation by the character $\chi_0^2\chi_1 \cdots \chi_n$.

Exercise 6.13: Let $\Gamma_n M \Gamma_n = \sqcup_i \Gamma_n M_i$, with

$$M_i = \begin{bmatrix} A_i & B_i \\ 0 & D_i \end{bmatrix} \quad \text{and} \quad D_i = \begin{bmatrix} p^{d_{i1}} & & * \\ & \ddots & \\ 0 & & p^{d_{in}} \end{bmatrix}.$$

Then the classical Satake p-parameters $\alpha_{0,p}, \cdots, \alpha_{n,p}$ satisfy

$$\lambda(T(M)) = \alpha_{0,p}^\delta \sum_i \prod_{j=1}^n (\alpha_{j,p}p^{-j})^{d_{ij}}$$

where δ is the valuation of $\mu(M)$. Because of the two slash actions considered, we get

$$T(M)\Phi_F = \mu(M)^{n(n+1-k)/2}\lambda(T(M))\Phi_F$$
$$= p^{\delta n(n+1-k)/2}\alpha_{0,p}^\delta \sum_i \prod_{j=1}^n (\alpha_{j,p}p^{-j})^{d_{ij}} \Phi_F.$$

Compare this to the formula for the Satake map to get

$$b_0 = p^{n(n+1)/4 - nk/2}\alpha_{0,p} \text{ and } b_i = \alpha_{i,p} \text{ for } i = 1, \cdots, n.$$

This completes proof of (i). Using the relation $\alpha_{0,p}^2 \alpha_{1,p} \cdots \alpha_{n,p} = p^{kn-n(n+1)/2}$ and the fact that the value of the central character at p is

$$b_0^2 b_1 \cdots b_n = p^{-kn+n(n+1)/2} \alpha_{0,p}^2 \alpha_{1,p} \cdots \alpha_{n,p} = 1.$$

Since the central character is unramified, it is determined by its value at p, and hence is trivial. This gives us part (ii) of the problem.

The relation $b_0 = p^{n(n+1)/4 - nk/2} \alpha_{0,p}$ and $b_i = \alpha_{i,p}$ for $i = 1, \cdots, n$ gives us the following relation between spin L-functions.

$$L(s, \pi_F, \text{spin}) = L(s - n(n+1)/4 + nk/2, F, \text{spin}).$$

For $n = 2$, we have

$$
\begin{aligned}
\Lambda(1 - s, \pi_F, \text{spin}) &= \Lambda(1 - s - 3/2 + k, F, \text{spin}) \\
&= (-1)^k \Lambda(2k - 2 - (1 - s - 3/2 + k), F, \text{spin}) \\
&= (-1)^k \Lambda(s - 3/2 + k, F, \text{spin}) \\
&= (-1)^k \Lambda(s, \pi_F, \text{spin}).
\end{aligned}
$$

Chapter 7

Exercise 7.1: Recall that W acts on the torus element $\text{diag}(a, b, ca^{-1}, cb^{-1})$ by exchanging a and b, or replacing a with ca^{-1} and b with cb^{-1}. There are eight elements in W corresponding to these actions on the torus. Suppose we consider a Weyl group element $w \in W$ which exchanges a and b and also replaces a with ca^{-1}. Then, we have

$$
\begin{aligned}
\chi^w(\text{diag}(a, b, ca^{-1}, cb^{-1})) &= \chi(\text{diag}(b, ca^{-1}, cb^{-1}, a)) = \chi_1(ca^{-1})\chi_2(b)\sigma(c) \\
&= \chi_2(b)\chi_1^{-1}(a)(\chi_1\sigma)(c).
\end{aligned}
$$

Hence $(\chi_1 \times \chi_2 \rtimes \sigma)^w = \chi_2 \times \chi_1^{-1} \rtimes \chi_1\sigma$. Similarly, one can calculate for other elements of W. In this way, we get the following eight isomorphic representations

$$
\begin{aligned}
\chi_1 \times \chi_2 \rtimes \sigma &\simeq \chi_2 \times \chi_1 \rtimes \sigma \simeq \chi_1^{-1} \times \chi_2 \rtimes \chi_1\sigma \simeq \chi_2 \times \chi_1^{-1} \rtimes \chi_1\sigma \\
&\simeq \chi_1 \times \chi_2^{-1} \rtimes \chi_2\sigma \simeq \chi_2^{-1} \times \chi_1 \rtimes \chi_2\sigma \simeq \chi_1^{-1} \times \chi_2^{-1} \rtimes \chi_1\chi_2\sigma \\
&\simeq \chi_2^{-1} \times \chi_1^{-1} \rtimes \chi_1\chi_2\sigma.
\end{aligned}
$$

Exercise 7.2: Recall that (π, V) is the space of all smooth $\mathbb{C}$-valued functions ψ on $GL_2(\mathbb{Q}_p)$ satisfying

$$\psi\left(\begin{bmatrix} a & \\ * & b \end{bmatrix} g\right) = |a^{-1}b|^{1/2}\eta(a)\eta^{-1}(b)\psi(g).$$

Here, we are inducing from the lower triangular matrices, and hence the roles of a and b are reversed. The representation $\tau \rtimes \nu^{1/2}$ consists of smooth functions f on G_p taking values in V satisfying

$$f\left(\begin{bmatrix} A & * \\ & c\,{}^tA^{-1} \end{bmatrix} g\right) = |\det(A)c^{-1}|^{3/2}|c|^{1/2}|\det(A)|^{-1/2}\pi(A)(f(g))$$

$$= |\det(A)c^{-1}|\pi(A)(f(g)).$$

Finally, the representation $\nu^{-1/2}\eta^{-1} \rtimes \nu^{-1/2}\eta \rtimes \nu^{1/2}$ consists of smooth $\mathbb{C}$-valued functions F on G_p satisfying

$$F\left(\begin{bmatrix} a & * & * & \\ * & b & * & * \\ & & ca^{-1} & * \\ & & & cb^{-1} \end{bmatrix} g\right) = |ab^2||c|^{-3/2}|a|^{-1/2}\eta(a)|b|^{-1/2}\eta^{-1}(b)|c|^{1/2}F(g)$$

$$= |a|^{1/2}|b|^{3/2}|c|^{-1}\eta(a)\eta^{-1}(b)F(g).$$

Define the map $L : \tau \rtimes \nu^{1/2} \to \nu^{-1/2}\eta^{-1} \rtimes \nu^{-1/2}\eta \rtimes \nu^{1/2}$ by $(Lf)(g) := \big(f(g)\big)(1_2)$. To check that the map is well defined, we compute the following:

$$(Lf)\left(\begin{bmatrix} a & * & * & \\ * & b & * & * \\ & & ca^{-1} & * \\ & & & cb^{-1} \end{bmatrix} g\right) = f\left(\begin{bmatrix} a & * & * & \\ * & b & * & * \\ & & ca^{-1} & * \\ & & & cb^{-1} \end{bmatrix} g\right)(1_2)$$

$$= |abc^{-1}|\left(\pi\left(\begin{bmatrix} a & \\ * & b \end{bmatrix}\right)(f(g))\right)(1_2) = |abc^{-1}|(f(g))\left(\begin{bmatrix} a & \\ * & b \end{bmatrix}\right)$$

$$= |abc^{-1}||a^{-1}b|^{1/2}\eta(a)\eta^{-1}(b)(f(g))(1_2) = |a|^{1/2}|b|^{3/2}|c|^{-1}\eta(a)\eta^{-1}(b)(Lf)(g),$$

as required. To check that the map L preserves the action of the group, we have

$$(g \cdot Lf)(g') = Lf(g'g) = \big(f(g'g)\big)(1_2) = \big((g \cdot f)(g')\big)(1_2) = (L(g \cdot f))(g').$$

Clearly, L is injective. One can construct a G_p-equivariant injective map $\tilde{L} : \nu^{-1/2}\eta^{-1} \rtimes \nu^{-1/2}\eta \rtimes \nu^{1/2} \to \tau \rtimes \nu^{1/2}$ that is the inverse of L by the following formula. Let $F \in \nu^{-1/2}\eta^{-1} \rtimes \nu^{-1/2}\eta \rtimes \nu^{1/2}$. For $g \in G_p$ and $A \in GL_2(\mathbb{Q}_p)$, define

$$((\tilde{L}F)(g))(A) := |\det A|^{-3/2} F\left(\begin{bmatrix} A & \\ & {}^tA^{-1} \end{bmatrix} g\right).$$

One can check that $\tilde{L}$ is well defined and preserve the action of G_p by following similar computations as above. Finally, one can check that $L \circ \tilde{L} = \tilde{L} \circ L = \mathrm{Id}$, the identity map.

Exercise 7.3: We have

$$W_v\left(\begin{bmatrix} 1 & y & * & \\ x & 1 & * & * \\ & & 1 & -x \\ & & & 1 \end{bmatrix} g\right) = \ell(\pi\left(\begin{bmatrix} 1 & y & * & \\ x & 1 & * & * \\ & & 1 & -x \\ & & & 1 \end{bmatrix} g\right)v) = \ell(\pi\left(\begin{bmatrix} 1 & y & * & \\ x & 1 & * & * \\ & & 1 & -x \\ & & & 1 \end{bmatrix}\right)\pi(g)v)$$

$$= \psi(c_1 x + c_2 y)\ell(\pi(g)v) = \psi(c_1 x + c_2 y) W_v(g).$$

Exercise 7.4: Define $\ell : \mathcal{W}(\pi, \psi_{c_1,c_2}) \to \mathbb{C}$ by

$$\ell(W) := W(1).$$

Then, we can check that

$$\ell(\pi(\begin{bmatrix} 1 & y & * & \\ x & 1 & * & * \\ & & 1 & -x \\ & & & 1 \end{bmatrix})W) = (\pi(\begin{bmatrix} 1 & y & * & \\ x & 1 & * & * \\ & & 1 & -x \\ & & & 1 \end{bmatrix})W)(1)$$

$$= W(\begin{bmatrix} 1 & y & * & \\ x & 1 & * & * \\ & & 1 & -x \\ & & & 1 \end{bmatrix}) = \psi(c_1 x + c_2 y) W(1)$$

$$= \psi(c_1 x + c_2 y)\ell(W).$$

Hence, ℓ is a Whittaker functional. To see that it is nonzero, take $0 \neq W \in \mathcal{W}(\pi, \psi_{c_1,c_2})$. Then there is a $g \in G_p$ such that $W(g) \neq 0$. This means that $(\pi(g)W)(1) \neq 0$, which gives us $\ell(\pi(g)W) \neq 0$.

Exercise 7.5: Let $0 \neq \ell \in \mathrm{Hom}_{N(\mathbb{Q}_p)}(\pi, \psi_{1,1})$. Define ℓ_{c_1,c_2} by

$$\ell_{c_1,c_2}(v) := \ell(\pi(\begin{bmatrix} c_2 & & & \\ & c_1 c_2 & & \\ & & 1 & \\ & & & c_1^{-1} \end{bmatrix})v).$$

Then $0 \neq \ell_{c_1,c_2} \in \mathrm{Hom}_{N(\mathbb{Q}_p)}(\pi, \psi_{c_1,c_2})$.

Exercise 7.6: This is direct computation.

Exercise 7.7: We have

$$
\eta_n
\begin{bmatrix}
\mathbb{Z}_p & p^n\mathbb{Z}_p & \mathbb{Z}_p & \mathbb{Z}_p \\
\mathbb{Z}_p & \mathbb{Z}_p & \mathbb{Z}_p & p^{-n}\mathbb{Z}_p \\
\mathbb{Z}_p & p^n\mathbb{Z}_p & \mathbb{Z}_p & \mathbb{Z}_p \\
p^n\mathbb{Z}_p & p^n\mathbb{Z}_p & p^n\mathbb{Z}_p & \mathbb{Z}_p
\end{bmatrix}
\eta_n^{-1}
=
\begin{bmatrix}
-p^n\mathbb{Z}_p & -p^n\mathbb{Z}_p & -p^n\mathbb{Z}_p & -\mathbb{Z}_p \\
\mathbb{Z}_p & p^n\mathbb{Z}_p & \mathbb{Z}_p & \mathbb{Z}_p \\
p^n\mathbb{Z}_p & p^n\mathbb{Z}_p & p^n\mathbb{Z}_p & \mathbb{Z}_p \\
-p^n\mathbb{Z}_p & -p^{2n}\mathbb{Z}_p & -p^n\mathbb{Z}_p & -p^n\mathbb{Z}_p
\end{bmatrix}
\eta_n^{-1}
$$

$$
=
\begin{bmatrix}
\mathbb{Z}_p & p^n\mathbb{Z}_p & \mathbb{Z}_p & \mathbb{Z}_p \\
\mathbb{Z}_p & \mathbb{Z}_p & \mathbb{Z}_p & p^{-n}\mathbb{Z}_p \\
\mathbb{Z}_p & p^n\mathbb{Z}_p & \mathbb{Z}_p & \mathbb{Z}_p \\
p^n\mathbb{Z}_p & p^n\mathbb{Z}_p & p^n\mathbb{Z}_p & \mathbb{Z}_p
\end{bmatrix}
$$

Hence, $\eta_n K(n)\eta_n^{-1} \subset K(n)$. Let $v \in V_\pi$ such that $\pi(k)v = v$ for all $k \in K(n)$. Then, for every $k \in K(n)$, we have

$$
\pi(k)\big(\pi(\eta_n)v\big) = \pi(\eta_n)\big(\pi(\eta_n^{-1}k\eta_n)v\big) = \pi(\eta_n)v,
$$

as required.

Exercise 7.8: Let $T = \mathrm{char}(K(n)gK(n))$ and let $K(n)gK(n) = \sqcup_i g_i K(n)$. Then, the action of T on $v \in V(n)$ is given by $\mathrm{vol}(K(n)) \sum_i \pi(g_i)v$. Hence, for any $k \in K(n)$, we have

$$
\pi(k)\big(Tv\big) = \pi(k)\mathrm{vol}(K(n)) \sum_i \pi(g_i)v = \mathrm{vol}(K(n)) \sum_i \pi(kg_i)v
$$

$$
= \mathrm{vol}(K(n)) \sum_i \pi(g_i)v = Tv
$$

since $\{kg_i\}$ gives another set of coset representatives for $K(n)gK(n)/K(n)$.

Exercise 7.9: We will use the defining property of the induced model for $\pi = \chi_1 \times \chi_2 \rtimes \sigma$, namely,

$$
f_0\left(
\begin{bmatrix}
a & * & * & \\
* b & * & & * \\
 & ca^{-1} & & * \\
 & & & cb^{-1}
\end{bmatrix}
k\right) = |ab^2||c|^{-3/2}\chi_1(b)\chi_2(a)\sigma(c)f_0(1),
$$

for $k \in K_p$. In particular, we have $f_0(1) \neq 0$. We also know that, if $K_p g K_p = \sqcup_i g_i K_p$ and $T = \mathrm{char}(K_p g K_p)$, then $(Tf_0)(g') = \sum_i f_0(g'g_i)$. Since $V_\pi^{K_p}$ is one dimensional, we know that $Tf_0 = \lambda(g) f_0$. To find λ we evaluate Tf_0 at 1 (using the non-vanishing of $f_0(1)$). Using the single coset decomposition given in the chapter notes, we get for $g = \mathrm{diag}(p, p, 1, 1)$

$$
\begin{aligned}
\lambda &= p^3|p^p||p|^{-3/2}\chi_1(p)\chi_2(p)\sigma(p) + p|p||p|^{-3/2}\chi_2(p)\sigma(p) \\
&\quad + |p|^{-3/2}\sigma(p) + p^2|p^2||p|^{-3/2}\chi_1(p)\sigma(p) \\
&= p^{3/2}(\chi_1(p)\chi_2(p)\sigma(p) + \chi_2(p)\sigma(p) + \sigma(p) + \chi_1(p)\sigma(p)) \\
&= p^{3/2}\sigma(p)(1 + \chi_1(p))(1 + \chi_2(p)).
\end{aligned}
$$

Note that we have $|p| = 1/p$. Using the single coset decomposition for $g = \mathrm{diag}(p, p^2, p, 1)$ given in the chapter notes, we get

$$
\begin{aligned}
\mu &= p^4|pp^4||p^2|^{-3/2}\chi_1^2(p)\chi_2(p)\sigma^2(p) + p^3|p^2p^2||p^2|^{-3/2}\chi_1(p)\chi_2^2(p)\sigma^2(p) \\
&\quad + p|p^2||p^2|^{-3/2}\chi_1(p)\sigma^2(p) + |p||p^2|^{-3/2}\chi_2(p)\sigma^2(p) \\
&\quad + (p-1)|pp^2||p^2|^{-3/2}\chi_1(p)\chi_2(p)\sigma^2(p) \\
&\quad + p(p-1)|pp^2||p^2|^{-3/2}\chi_1(p)\chi_2(p)\sigma^2(p) \\
&= p^2(\chi_1(p) + \chi_1^{-1}(p) + \chi_2(p) + \chi_2^{-1}(p) + 1 - p^{-2}).
\end{aligned}
$$

We have used $\chi_1\chi_2\sigma^2 = 1$ here.

Exercise 7.11: This is direct computation.

Chapter 8

Exercise 8.3: We see that $\xi_S^2 = \frac{d}{4}1_2$. Hence, $F(\xi)$ is a quadratic extension of $\mathbb{Q}$. Once again, the relation $\xi_S^2 = \frac{d}{4}1_2$ tells us that the map is an isomorphism from $F(\xi)$ to $L = \mathbb{Q}(\sqrt{d})$. The details can easily be checked. Let $g = \begin{bmatrix} u & v \\ w & z \end{bmatrix} \in GL_2(\mathbb{Q})$. Then, ${}^tgSg = \det(g)S$ can be rewritten as ${}^tgS = \det(g)Sg^{-1}$. This gives us

$$
\begin{bmatrix} u & w \\ v & z \end{bmatrix}\begin{bmatrix} a & b/2 \\ b/2 & c \end{bmatrix} = \begin{bmatrix} a & b/2 \\ b/2 & c \end{bmatrix}\begin{bmatrix} z & -v \\ -w & u \end{bmatrix}.
$$

Comparing the components of the matrices on both sides, we get

$$
ua + bw/2 = az - bw/2, \quad ub/2 + wc = ub/2 - av, \quad bv/2 + zc = -bv/2 + cu.
$$

Hence, we get $wc = -av$ and $(u - z)c = bv$. If we set $wc = -av = -acy$ and $u + z = 2x$, then we get

$$g = \begin{bmatrix} x + by/2 & yc \\ -ya & x - by/2 \end{bmatrix},$$

which is an invertible element of $F(\xi)$. Hence, we get $T = F(\xi)^\times$.

Exercise 8.4: Let $t = \begin{bmatrix} g & \\ & \det(g)\, {}^tg^{-1} \end{bmatrix}$ and $u = \begin{bmatrix} 1 & X \\ & 1 \end{bmatrix}$. Then

$$t^{-1}ut = \begin{bmatrix} 1 & \det(g)g^{-1}X\,{}^tg^{-1} \\ & 1 \end{bmatrix}.$$

Hence, we have

$$\theta(t^{-1}ut) = \psi(\mathrm{Tr}(S\det(g)g^{-1}X\,{}^tg^{-1})) = \psi(\mathrm{Tr}(\det(g)\,{}^tg^{-1}Sg^{-1}X))$$
$$= \psi(\mathrm{Tr}(SX)) = \theta(u).$$

Here, we have used the definition of T. Hence, we have

$$(\Lambda \otimes \theta)(ut) = (\Lambda \otimes \theta)(tt^{-1}ut) = \Lambda(t)\theta(t^{-1}ut) = \Lambda(t)\theta(u) = (\Lambda \otimes \theta)(tu).$$

Exercise 8.5: This follows from a straightforward change of variable in the integral defining B_ϕ.

Exercise 8.7: This follows exactly as in the case of the Whittaker model discussed in the previous chapter.

Exercise 8.8: Follows from the definitions.

Exercise 8.9: Since Φ_F is right invariant under $K_0 = \prod_p K_p$, we can see that B_{Φ_F} is also right invariant under $K_0 = \prod_p K_p$. Since B_p is the restriction of B_{Φ_F} to G_p, we see that B_p is right invariant under K_p.

Exercise 8.10: We have

$$\mathrm{GSp}_4(\mathbb{Q}_p) = \bigsqcup_{l \in \mathbb{Z}, m \geq 0} R(\mathbb{Q}_p)h_p(l, m)K_p.$$

Write any $g \in G_p$ as $g = rh_p(l, m)k$ with $r \in R(\mathbb{Q}_p)$ and $k \in K_p$. Then, we have

$$B_p(g) = B_p(rh_p(l, m)k) = (\Lambda \otimes \theta)(r)B_p(h_p(l, m)).$$

This shows that B_p is completely determined by its values on $h_p(l, m)$ for $l \in \mathbb{Z}$, $m \geq 0$. Next, take $X = \begin{bmatrix} x & y \\ y & z \end{bmatrix} \in M_2(\mathbb{Z}_p)$. We have

$$B_p(h_p(l, m)) = B_p(h_p(l, m)\begin{bmatrix} 1 & X \\ & 1 \end{bmatrix}) = B_p(h_p(l, m)\begin{bmatrix} 1 & X \\ & 1 \end{bmatrix}h_p(l, m)^{-1}h_p(l, m))$$

$$= B_p(\begin{bmatrix} 1 & p^{2m+l}x & p^{m+l}y \\ & 1 & p^{m+l}y & p^l z \\ & & 1 \\ & & & 1 \end{bmatrix}h_p(l, m))$$

$$= \psi_p(\mathrm{Tr}(S\begin{bmatrix} p^{2m+l}x & p^{m+l}y \\ p^{m+l}y & p^l z \end{bmatrix}))B_p(h_p(l, m))$$

$$= \psi_p(p^{2m+l}ax + p^{m+l}by + p^l cz)B_p(h_p(l, m)).$$

Now, suppose $l < 0$. Since $c \in \mathbb{Z}_p^\times$ and conductor of ψ_p is $\mathbb{Z}_p$, we can find a $z \in \mathbb{Z}_p$ such that $\psi_p(p^l cz) \neq 1$. Take such a z and set $x = y = 0$. Then, we get

$$B_p(h_p(l, m)) = \psi_p(p^l cz)B_p(h_p(l, m)),$$

which implies that $B_p(h_p(l, m)) = 0$.

Exercise 8.12: Let us drop the subscript p. The above recurrence relation is equivalent to the following identity between generating series,

$$\sum_{l,m \geq 0} B(h(l, m))x^m y^l = \sum_{l,m \geq 0} \sum_{i=0}^{l} p^{-i} B(h(0, l + m - i))x^m y^l. \tag{D.1}$$

By Sugano's formula, the left-hand side equals

$$\mathrm{LHS} = \frac{H(x, y)}{P(x)Q(y)},$$

where H, P, Q are defined as in Theorem 8.11. For the right-hand side of (D.1), we calculate

$$\text{RHS} = \sum_{m=0}^{\infty} \sum_{l=0}^{\infty} \sum_{i=0}^{l} p^{-i} B(h(0, l+m-i)) x^m y^l$$

$$= \sum_{m=0}^{\infty} \sum_{i=0}^{\infty} \sum_{l=i}^{\infty} p^{-i} B(h(0, l+m-i)) x^m y^l$$

$$= \sum_{m=0}^{\infty} \sum_{i=0}^{\infty} \sum_{l=0}^{\infty} p^{-i} B(h(0, l+m)) x^m y^{l+i}$$

$$= \frac{1}{1-p^{-1}y} \sum_{m=0}^{\infty} \sum_{l=0}^{\infty} B(h(0, l+m)) x^m y^l$$

$$= \frac{1}{1-p^{-1}y} \sum_{j=0}^{\infty} \sum_{l+m=j} B(h(0, j)) x^m y^l$$

$$= \frac{1}{1-p^{-1}y} \sum_{j=0}^{\infty} B(h(0, j)) \frac{x^{j+1} - y^{j+1}}{x - y}$$

$$= \frac{1}{(1-p^{-1}y)(x-y)} \left(x \sum_{j=0}^{\infty} B(h(0, j)) x^j - y \sum_{j=0}^{\infty} B(h(0, j)) y^j \right)$$

$$= \frac{1}{(1-p^{-1}y)(x-y)} \left(x \frac{H(x, 0)}{P(x) Q(0)} - y \frac{H(y, 0)}{P(y) Q(0)} \right)$$

$$= \frac{1}{(1-p^{-1}y)(x-y)} \left(x \frac{H(x, 0)}{P(x)} - y \frac{H(y, 0)}{P(y)} \right).$$

Hence, (D.1) is equivalent to

$$(1 - p^{-1}y)(x - y) H(x, y) P(y) - Q(y) \Big(x P(y) H(x, 0) - y P(x) H(y, 0) \Big) = 0.$$
(D.2)

If one of the Satake parameters is $p^{\pm 1/2}$, then one can verify that (D.2) is satisfied.

Exercise 8.13: Follows from Theorem 8.11, and the definitions of γ_i and the local L-functions.

Chapter 9

Exercise 9.4: Since π_F is unramified at all finite primes, so is Π_4. Since Π_4 is assumed to be a constituent of a globally induced representation from a proper parabolic subgroup of GL_4, the inducing data has to be unramified at all finite primes as well. The four possibilities of proper parabolics of GL_4 give us the following four possible factorizations of the L-function $L(s, \Pi_4)$.

$$L(s, \chi_1)L(s, \chi_2)L(s, \sigma), \quad L(s, \sigma_1)L(s, \sigma_2), \quad L(s, \chi)L(s, \tau)$$
$$L(s, \chi_1)L(s, \chi_2)L(s, \chi_3)L(s, \chi_4)$$

Here, the χ's are characters of $\mathbb{A}^\times$, σ's are cuspidal representations of $GL_2(\mathbb{A})$ and τ is a cuspidal representation of $GL_3(\mathbb{A})$. Note that these representations are unramified at all finite primes. In particular, the characters are all trivial. Hence, whenever we have a L-factor corresponding to a character, it contributes a pole at $s = 1$, which is not canceled by any zeros of the remaining terms. Note that $L(s, \Pi_4)$ is entire since F is not a Saito–Kurokawa lift. So, the only remaining possibility is $L(s, \sigma_1)L(s, \sigma_2)$. For this case, consider the L-function $L(s, \Pi_4 \times \tilde{\sigma}_1) = L(s, \sigma_1 \times \tilde{\sigma}_1)L(s, \sigma_2 \times \tilde{\sigma}_1)$. Since $\tilde{\sigma}_1$ is unramified everywhere, it satisfies the hypothesis that it is unramified for $p|d$. Hence, $L(s, \Pi_4 \times \tilde{\sigma}_1)$ is entire. But we know that $L(s, \sigma_1 \times \tilde{\sigma}_1)$ has a pole at $s = 1$ which is not canceled by a zero of the other term. Hence, Π_4 is cuspidal.

Exercise 9.5: For part (i), we have

$$L(s, \tau_f, \Lambda^2) = \prod_p (1 - \alpha_p \bar{\alpha}_p p^{-s})^{-1} = \prod_p (1 - p^{-s})^{-1} = \zeta(s)$$

and

$$L(s, \tau_f, \text{Sym}^2) = \prod_p \left((1 - \alpha_p^2 p^{-s})(1 - p^{-s})(1 - \bar{\alpha}_p^2 p^{-s}) \right)^{-1}.$$

For part (ii), $L(s, \Pi_4, \Lambda^2)$

$$= \prod_p \left((1 - b_0^2 b_1 p^{-s})(1 - b_0^2 b_2 p^{-s})(1 - b_0^2 b_1 b_2 p^{-s}) \right.$$
$$\left. \times (1 - b_0^2 b_1 b_2 p^{-s})(1 - b_0^2 b_1^2 b_2 p^{-s})(1 - b_0^2 b_1 b_2^2 p^{-s}) \right)^{-1}$$
$$= \prod_p \left((1 - b_2^{-1} p^{-s})(1 - b_1^{-1} p^{-s})(1 - p^{-s})(1 - p^{-s})(1 - b_1 p^{-s})(1 - b_2 p^{-s}) \right)^{-1}$$
$$= \prod_p \left((1 - b_2^{-1} p^{-s})(1 - b_1^{-1} p^{-s})(1 - p^{-s})(1 - b_1 p^{-s})(1 - b_2 p^{-s}) \right)^{-1} \times \prod_p (1 - p^{-s})^{-1}$$
$$= L(s, \pi_F, \text{std})\zeta(s),$$

as required. Here, we have used $b_0^2 b_1 b_2 = 1$.

Exercise 9.6: We have to consider $\Gamma(s)$ and $\Gamma(\ell - s)$. From the first one, we see that we have a pole for all nonpositive integers. And from the second one, we get poles for all integers greater than or equal to ℓ. The only remaining ones are integers m satisfying $0 < m < \ell$.

Exercise 9.7: The gamma factors involved in this case are $\Gamma(s + \frac{k}{2})$, $\Gamma(s + \frac{k}{2} - 1)$ and $\Gamma(s + \frac{3k}{2})$. Hence, we want to find integers m such that none of the three gamma functions above have a pole at $s = m$. In addition, $\Gamma(1 - s + \frac{k}{2})$, $\Gamma(1 - s + \frac{k}{2} - 1)$ and $\Gamma(1 - s + \frac{3k}{2})$ do not have a pole at $s = m$. We can see that $\Gamma(s + \frac{k}{2})\Gamma(s + \frac{k}{2} - 1)\Gamma(s + \frac{3k}{2})$ has poles exactly at integers $m \leq -k/2 + 1$. Similarly, $\Gamma(1 - s + \frac{k}{2})\Gamma(-s + \frac{k}{2})\Gamma(1 - s + \frac{3k}{2})$ has poles exactly at integers $m \geq k/2$. Hence, the set of critical points is the set of all integers in the interval $[-\frac{k}{2} + 2, \frac{k}{2} - 1]$.

Chapter 10

Exercise 10.1: By direct computation.

Exercise 10.2: Substitute the definition of the Eisenstein series in the integral, and use the relevant double and single coset decompositions to get

$$
Z(s; f, \phi)(g) = \int_{\mathrm{Sp}_{2n}(F) \backslash g \cdot \mathrm{Sp}_{2n}(\mathbb{A})} E((h, g), s, f) \phi(h) \, dh
$$

$$
= \int_{\mathrm{Sp}_{2n}(F) \backslash g \cdot \mathrm{Sp}_{2n}(\mathbb{A})} \sum_{\gamma \in P_{4n}(F) \backslash G_{4n}(F)} f(\gamma(h, g), s) \phi(h) \, dh
$$

$$
= \int_{\mathrm{Sp}_{2n}(F) \backslash g \cdot \mathrm{Sp}_{2n}(\mathbb{A})} \sum_{r=0}^{n} \sum_{i} f(Q_r \gamma_i^{(r)}(h, g), s) \phi(h) \, dh
$$

$$
= \sum_{r=0}^{n} Z_r(s; f, \phi).
$$

Exercise 10.4: Suppose $p_1 Q_n \gamma_1 = p_2 Q_n \gamma_2$ with $p_1, p_2 \in P_{4n}(F)$ and $\gamma_1, \gamma_2 \in \Gamma_{4n}(\mathfrak{p}^m)$. Then $Q_n^{-1} p_2^{-1} p_1 Q_n \in \Gamma_{4n}(\mathfrak{p}^m)$. Hence, we have $p_2^{-1} p_1 \in \Gamma_{4n}(\mathfrak{p}^m)$. Now, well-definedness of f follows from the right-$\Gamma_{4n}(\mathfrak{p}^m)$ invariance of f, and the fact that both χ and $\delta_{P_{4n}}$ are invariant under $\Gamma_{4n}(\mathfrak{p}^m)$.

Exercise 10.5: The leftmost zero in the numerator is $z = k - 1 - j - 2i$ for $j = n$ and $i = (k - k_n)/2 - 1$, i.e., $z = k_n - n + 1$. Hence, if $t \leq k_n - n$ then we do not have a zero. Also, for $t \geq 0$, the denominators are all positive integers. Hence, we get the result.

Exercise 10.9: For a prime $p \nmid M$, let the Satake p-parameters of π be given by $\alpha_{0,p}, \alpha_{1,p}, \alpha_{2,p}$. The L-function condition implies that

$$\{\alpha_{0,p}, \alpha_{0,p}\alpha_{1,p}, \alpha_{0,p}\alpha_{2,p}, \alpha_{0,p}\alpha_{1,p}\alpha_{2,p}\} = \{\alpha_p^3, \alpha_p, \alpha_p^{-1}, \alpha_p^{-3}\}.$$

Using $\alpha_{0,p}^2\alpha_{1,p}\alpha_{2,p} = 1$, we see that

$$\{\alpha_{1,p}, \alpha_{1,p}^{-1}, \alpha_{2,p}, \alpha_{2,p}^{-1}\} = \{\alpha_p^2, \alpha_p^{-2}, \alpha_p^4, \alpha_p^{-4}\}.$$

This gives us the result.

References

1. Mahesh Agarwal and Jim Brown. Saito-Kurokawa lifts of square-free level. *Kyoto J. Math.*, 55(3):641–662, 2015.
2. A. N. Andrianov. Euler products that correspond to Siegel's modular forms of genus 2. *Uspehi Mat. Nauk*, 29(3 (177)):43–110, 1974.
3. A. N. Andrianov. Multiplicative arithmetic of Siegel's modular forms. *Uspekhi Mat. Nauk*, 34(1(205)):67–135, 1979.
4. Anatoli Andrianov. *Introduction to Siegel modular forms and Dirichlet series*. Universitext. Springer, New York, 2009.
5. James Arthur. *The endoscopic classification of representations*, volume 61 of *American Mathematical Society Colloquium Publications*. American Mathematical Society, Providence, RI, 2013. Orthogonal and symplectic groups.
6. Mahdi Asgari and Ralf Schmidt. Siegel modular forms and representations. *Manuscripta Math.*, 104(2):173–200, 2001.
7. S. Böcherer. Ein Rationalitätssatz für formale Heckereihen zur Siegelschen Modulgruppe. *Abh. Math. Sem. Univ. Hamburg*, 56:35–47, 1986.
8. S. Böcherer and R. Schulze-Pillot. Siegel modular forms and theta series attached to quaternion algebras. II. *Nagoya Math. J.*, 147:71–106, 1997. With errata to: "Siegel modular forms and theta series attached to quaternion algebras" [Nagoya Math. J. 121 (1991), 35–96; MR1096467 (92f:11066)].
9. Siegfried Böcherer. über die Funktionalgleichung automorpher L-Funktionen zur Siegelschen Modulgruppe. *J. Reine Angew. Math.*, 362:146–168, 1985.
10. Siegfried Böcherer. Bemerkungen über die Dirichletreihen von Koecher und Maass. *Mathematica Gottingensis*, 68:36 pp., 1986.
11. Siegfried Böcherer and Soumya Das. Characterization of Siegel cusp forms by the growth of their Fourier coefficients. *Math. Ann.*, 359(1–2):169–188, 2014.
12. Siegfried Böcherer and Bernhard E. Heim. Critical values of L-functions on $GSp_2 \times GL_2$. *Math. Z.*, 254(3):485–503, 2006.
13. Siegfried Böcherer and Rainer Schulze-Pillot. Siegel modular forms and theta series attached to quaternion algebras. *Nagoya Math. J.*, 121:35–96, 1991.
14. Stefan Breulmann. On Hecke eigenforms in the Maaß space. *Math. Z.*, 232(3):527–530, 1999.
15. Jim Brown. The first negative Hecke eigenvalue of genus 2 Siegel cuspforms with level $N \geq 1$. *Int. J. Number Theory*, 6(4):857–867, 2010.
16. Armand Brumer and Kenneth Kramer. Paramodular abelian varieties of odd conductor. *Trans. Amer. Math. Soc.*, 366(5):2463–2516, 2014.

© Springer Nature Switzerland AG 2019
A. Pitale, *Siegel Modular Forms*, Lecture Notes in Mathematics 2240,
https://doi.org/10.1007/978-3-030-15675-6

17. Daniel Bump. *Automorphic forms and representations*, volume 55 of *Cambridge Studies in Advanced Mathematics*. Cambridge University Press, Cambridge, 1997.

18. YoungJu Choie, Sanoli Gun, and Winfried Kohnen. An explicit bound for the first sign change of the Fourier coefficients of a Siegel cusp form. *Int. Math. Res. Not. IMRN*, (12):3782–3792, 2015.

19. YoungJu Choie and Winfried Kohnen. The first sign change of Fourier coefficients of cusp forms. *Amer. J. Math.*, 131(2):517–543, 2009.

20. J. W. Cogdell and I. I. Piatetski-Shapiro. A converse theorem for GL$_4$. *Math. Res. Lett.*, 3(1):67–76, 1996.

21. P. Deligne. Valeurs de fonctions L et périodes d'intégrales. In *Automorphic forms, representations and L-functions (Proc. Sympos. Pure Math., Oregon State Univ., Corvallis, Ore., 1977), Part 2*, Proc. Sympos. Pure Math., XXXIII, pages 313–346. Amer. Math. Soc., Providence, R.I., 1979. With an appendix by N. Koblitz and A. Ogus.

22. Pierre Deligne. La conjecture de Weil. I. *Inst. Hautes Études Sci. Publ. Math.*, (43):273–307, 1974.

23. Martin Dickson, Ameya Pitale, Abhishek Saha, and Ralf Schmidt. Explicit refinements of Böcherer's conjecture for Siegel modular forms of squarefree level. *to appear in J. Math. Soc. Japan*, 2018.

24. W. Duke and Ö. Imamoglu. A converse theorem and the Saito-Kurokawa lift. *Internat. Math. Res. Notices*, (7):347–355, 1996.

25. Martin Eichler and Don Zagier. *The theory of Jacobi forms*, volume 55 of *Progress in Mathematics*. Birkhäuser Boston, Inc., Boston, MA, 1985.

26. S. A. Evdokimov. Characterization of the Maass space of Siegel modular cusp forms of genus 2. *Mat. Sb. (N.S.)*, 112(154)(1(5)):133–142, 144, 1980.

27. David Farmer, Nathan Ryan, Ameya Pitale, and Ralf Schmidt. Multiplicity one for L-functions and applications. *preprint*, 2018.

28. E. Freitag. *Siegelsche Modulfunktionen*, volume 254 of *Grundlehren der Mathematischen Wissenschaften [Fundamental Principles of Mathematical Sciences]*. Springer-Verlag, Berlin, 1983.

29. William Fulton and Joe Harris. *Representation theory*, volume 129 of *Graduate Texts in Mathematics*. Springer-Verlag, New York, 1991. A first course, Readings in Mathematics.

30. Masaaki Furusawa. On L-functions for GSp(4) × GL(2) and their special values. *J. Reine Angew. Math.*, 438:187–218, 1993.

31. Masaaki Furusawa and Kazuki Morimoto. On special values of certain L-functions, II. *Amer. J. Math.*, 138(4):1117–1166, 2016.

32. Masaaki Furusawa and Kazuki Morimoto. On special Bessel periods and the Gross-Prasad conjecture for SO(2n + 1) × SO(2). *Math. Ann.*, 368(1-2):561–586, 2017.

33. Masaaki Furusawa and Kazuki Morimoto. Refined global Gross-Prasad conjecture on special Bessel periods and Böcherer's conjecture. *preprint*, 2017.

34. Wee Teck Gan and Shuichiro Takeda. The local Langlands conjecture for GSp(4). *Ann. of Math. (2)*, 173(3):1841–1882, 2011.

35. Paul B. Garrett. On the arithmetic of Siegel-Hilbert cuspforms: Petersson inner products and Fourier coefficients. *Invent. Math.*, 107(3):453–481, 1992.

36. Paul B. Garrett and Michael Harris. Special values of triple product L-functions. *Amer. J. Math.*, 115(1):161–240, 1993.

37. David Ginzburg, Stephen Rallis, and David Soudry. Generic automorphic forms on SO(2n + 1): functorial lift to GL(2n), endoscopy, and base change. *Internat. Math. Res. Notices*, (14):729–764, 2001.

38. Michael Harris. Eisenstein series on Shimura varieties. *Ann. of Math. (2)*, 119(1):59–94, 1984.

39. T. Ibukiyama and H. Katsurada. An Atkin-Lehner type theorem on Siegel modular forms and primitive Fourier coefficients. In *Geometry and analysis of automorphic forms of several variables*, volume 7 of *Ser. Number Theory Appl.*, pages 196–210. World Sci. Publ., Hackensack, NJ, 2012.

40. Tomoyoshi Ibukiyama and Hidenori Katsurada. Exact critical values of the symmetric fourth L function and vector valued Siegel modular forms. *J. Math. Soc. Japan*, 66(1):139–160, 2014.
41. Atsushi Ichino. A regularized Siegel-Weil formula for unitary groups. *Math. Z.*, 247(2):241–277, 2004.
42. Jun-ichi Igusa. Modular forms and projective invariants. *Amer. J. Math.*, 89:817–855, 1967.
43. Kaori Imai. Generalization of Hecke's correspondence to Siegel modular forms. *Amer. J. Math.*, 102(5):903–936, 1980.
44. S. Jesgarz. Vorzeichenwechsel von fourierkoeffizienten von siegelschen spitzenformen. *Diploma Thesis (unpublished), University of Heidelberg*, 2008.
45. Henry H. Kim. Functoriality for the exterior square of GL_4 and the symmetric fourth of GL_2. *J. Amer. Math. Soc.*, 16(1):139–183, 2003. With appendix 1 by Dinakar Ramakrishnan and appendix 2 by Kim and Peter Sarnak.
46. Helmut Klingen. *Introductory lectures on Siegel modular forms*, volume 20 of *Cambridge Studies in Advanced Mathematics*. Cambridge University Press, Cambridge, 1990.
47. Martin Kneser. Strong approximation. In *Algebraic Groups and Discontinuous Subgroups (Proc. Sympos. Pure Math., Boulder, Colo., 1965)*, pages 187–196. Amer. Math. Soc., Providence, R.I., 1966.
48. Winfried Kohnen. Modular forms of half-integral weight on $\Gamma_0(4)$. *Math. Ann.*, 248(3):249–266, 1980.
49. Winfried Kohnen. Sign changes of Hecke eigenvalues of Siegel cusp forms of genus two. *Proc. Amer. Math. Soc.*, 135(4):997–999, 2007.
50. Winfried Kohnen. On certain generalized modular forms. *Funct. Approx. Comment. Math.*, 43(part 1):23–29, 2010.
51. Winfried Kohnen and Yves Martin. A characterization of degree two Siegel cusp forms by the growth of their Fourier coefficients. *Forum Math.*, 26(5):1323–1331, 2014.
52. Winfried Kohnen and Jyoti Sengupta. The first negative Hecke eigenvalue of a Siegel cusp form of genus two. *Acta Arith.*, 129(1):53–62, 2007.
53. Noritomo Kozima. On special values of standard L-functions attached to vector valued Siegel modular forms. *Kodai Math. J.*, 23(2):255–265, 2000.
54. Aloys Krieg and Martin Raum. The functional equation for the twisted spinor L-series of genus 2. *Abh. Math. Semin. Univ. Hambg.*, 83(1):29–52, 2013.
55. Edmund Landau. Über einen Satz von Tschebyschef. *Math. Ann.*, 61(4):527–550, 1906.
56. Erez M. Lapid. On the nonnegativity of Rankin-Selberg L-functions at the center of symmetry. *Int. Math. Res. Not.*, (2):65–75, 2003.
57. Gérard Laumon. Fonctions zêtas des variétés de Siegel de dimension trois. *Astérisque*, (302):1–66, 2005. Formes automorphes. II. Le cas du groupe $GSp(4)$.
58. Jian-Shu Li. Nonexistence of singular cusp forms. *Compositio Math.*, 83(1):43–51, 1992.
59. Hans Maass. *Siegel's modular forms and Dirichlet series*. Lecture Notes in Mathematics, Vol. 216. Springer-Verlag, Berlin-New York, 1971. Dedicated to the last great representative of a passing epoch. Carl Ludwig Siegel on the occasion of his seventy-fifth birthday.
60. Hans Maass. über eine Spezialschar von Modulformen zweiten Grades. *Invent. Math.*, 52(1):95–104, 1979.
61. Hans Maass. über eine Spezialschar von Modulformen zweiten Grades. II. *Invent. Math.*, 53(3):249–253, 1979.
62. Toshitsune Miyake. *Modular forms*. Springer Monographs in Mathematics. Springer-Verlag, Berlin, english edition, 2006. Translated from the 1976 Japanese original by Yoshitaka Maeda.
63. Kazuki Morimoto. On L-functions for quaternion unitary groups of degree 2 and GL(2) (with an appendix by M. Furusawa and A. Ichino). *Int. Math. Res. Not. IMRN*, (7):1729–1832, 2014.
64. Hiro-aki Narita, Ameya Pitale, and Ralf Schmidt. Irreducibility criteria for local and global representations. *Proc. Amer. Math. Soc.*, 141(1):55–63, 2013.
65. Takayuki Oda. On modular forms associated with indefinite quadratic forms of signature $(2, n-2)$. *Math. Ann.*, 231(2):97–144, 1977/78.

66. I. Piatetski-Shapiro and S. Rallis. *L*-functions of automorphic forms on simple classical groups. In *Modular forms (Durham, 1983)*, Ellis Horwood Ser. Math. Appl.: Statist. Oper. Res., pages 251–261. Horwood, Chichester, 1984.

67. Ameya Pitale. Classical interpretation of the Ramanujan conjecture for Siegel cusp forms of genus *n*. *Manuscripta Math.*, 130(2):225–231, 2009.

68. Ameya Pitale. Steinberg representation of GSp(4): Bessel models and integral representation of *L*-functions. *Pacific J. Math.*, 250(2):365–406, 2011.

69. Ameya Pitale, Abhishek Saha, and Ralf Schmidt. Transfer of Siegel cusp forms of degree 2. *Mem. Amer. Math. Soc.*, 232(1090):vi+107, 2014.

70. Ameya Pitale, Abhishek Saha, and Ralf Schmidt. Local and global Maass relations. *Math. Z.*, 287(1-2):655–677, 2017.

71. Ameya Pitale, Abhishek Saha, and Ralf Schmidt. Lowest weight modules of Sp(4, ℝ) and nearly holomorphic siegel modular forms. *preprint*, 2018.

72. Ameya Pitale, Abhishek Saha, and Ralf Schmidt. On the standard l-function for $GSp_{2n} \times GL_1$ and algebraicity of symmetric fourth *L*-values for GL_2. *preprint*, 2018.

73. Ameya Pitale and Ralf Schmidt. Sign changes of Hecke eigenvalues of Siegel cusp forms of degree 2. *Proc. Amer. Math. Soc.*, 136(11):3831–3838, 2008.

74. Ameya Pitale and Ralf Schmidt. Bessel models for lowest weight representations of GSp(4, ℝ). *Int. Math. Res. Not. IMRN*, (7):1159–1212, 2009.

75. Ameya Pitale and Ralf Schmidt. Integral representation for *L*-functions for $GSp_4 \times GL_2$. *J. Number Theory*, 129(6):1272–1324, 2009.

76. Aaron Pollack. The spin L-function on GSp_6 for Siegel modular forms. *Compositio Mathematica*, 153(7):1391–1432, 2017.

77. Cris Poor and David S. Yuen. Linear dependence among Siegel modular forms. *Math. Ann.*, 318(2):205–234, 2000.

78. Dipendra Prasad and Ramin Takloo-Bighash. Bessel models for GSp(4). *J. Reine Angew. Math.*, 655:189–243, 2011.

79. A. Raghuram. On the special values of certain Rankin-Selberg *L*-functions and applications to odd symmetric power *L*-functions of modular forms. *Int. Math. Res. Not. IMRN*, (2):334–372, 2010.

80. Dinakar Ramakrishnan and Freydoon Shahidi. Siegel modular forms of genus 2 attached to elliptic curves. *Math. Res. Lett.*, 14(2):315–332, 2007.

81. Brooks Roberts and Ralf Schmidt. *Local newforms for GSp(4)*, volume 1918 of *Lecture Notes in Mathematics*. Springer, Berlin, 2007.

82. Brooks Roberts and Ralf Schmidt. Some results on Bessel functionals for GSp(4). *Doc. Math.*, 21:467–553, 2016.

83. François Rodier. Whittaker models for admissible representations of reductive *p*-adic split groups. In *Harmonic analysis on homogeneous spaces (Proc. Sympos. Pure Math., Williams Coll., Williamstown, Mass., 1972)*. Proc. Sympos. Pure Math., XXVI, pages 425–430. Amer. Math. Soc. Providence, R.I., 1973.

84. Emmanuel Royer, Jyoti Sengupta, and Jie Wu. Non-vanishing and sign changes of Hecke eigenvalues for Siegel cusp forms of genus two. *Ramanujan J.*, 39(1):179–199, 2016. With an appendix by E. Kowalski and A. Saha.

85. Abhishek Saha. *L*-functions for holomorphic forms on GSp(4) × GL(2) and their special values. *Int. Math. Res. Not. IMRN*, (10):1773–1837, 2009.

86. Abhishek Saha. Pullbacks of Eisenstein series from GU(3, 3) and critical *L*-values for GSp(4) × GL(2). *Pacific J. Math.*, 246(2):435–486, 2010.

87. Abhishek Saha. Determination of modular forms by fundamental Fourier coefficients. In *Automorphic representations and L-functions*, volume 22 of *Tata Inst. Fundam. Res. Stud. Math.*, pages 501–519. Tata Inst. Fund. Res., Mumbai, 2013.

88. Abhishek Saha. Siegel cusp forms of degree 2 are determined by their fundamental Fourier coefficients. *Math. Ann.*, 355(1):363–380, 2013.

89. Abhishek Saha and Ralf Schmidt. Yoshida lifts and simultaneous non-vanishing of dihedral twists of modular *L*-functions. *J. Lond. Math. Soc. (2)*, 88(1):251–270, 2013.

90. Paul J. Sally, Jr. and Marko Tadić. Induced representations and classifications for GSp(2, *F*) and Sp(2, *F*). *Mém. Soc. Math. France (N.S.)*, (52):75–133, 1993.

91. Peter Sarnak. Notes on the generalized Ramanujan conjectures. In *Harmonic analysis, the trace formula, and Shimura varieties*, volume 4 of *Clay Math. Proc.*, pages 659–685. Amer. Math. Soc., Providence, RI, 2005.

92. Ralf Schmidt. Iwahori-spherical representations of GSp(4) and Siegel modular forms of degree 2 with square-free level. *J. Math. Soc. Japan*, 57(1):259–293, 2005.

93. Ralf Schmidt. Packet structure and paramodular forms. *Trans. Amer. Math. Soc.*, 370(5):3085–3112, 2018.

94. Ralf Schmidt and Alok Shukla. On klingen eisenstein series with level in degree two. *to appear in The Journal of the Ramanujan Mathematical Society*, 2018.

95. Goro Shimura. On the periods of modular forms. *Math. Ann.*, 229(3):211–221, 1977.

96. Goro Shimura. Convergence of zeta functions on symplectic and metaplectic groups. *Duke Math. J.*, 82(2):327–347, 1996.

97. Goro Shimura. *Euler products and Eisenstein series*, volume 93 of *CBMS Regional Conference Series in Mathematics*. Published for the Conference Board of the Mathematical Sciences, Washington, DC; by the American Mathematical Society, Providence, RI, 1997.

98. Goro Shimura. *Arithmeticity in the theory of automorphic forms*, volume 82 of *Mathematical Surveys and Monographs*. American Mathematical Society, Providence, RI, 2000.

99. Carl Ludwig Siegel. Einführung in die Theorie der Modulfunktionen *n*-ten Grades. *Math. Ann.*, 116:617–657, 1939.

100. Jacob Sturm. Special values of zeta functions, and Eisenstein series of half integral weight. *Amer. J. Math.*, 102(2):219–240, 1980.

101. Takashi Sugano. On holomorphic cusp forms on quaternion unitary groups of degree 2. *J. Fac. Sci. Univ. Tokyo Sect. IA Math.*, 31(3):521–568, 1985.

102. Ramin Takloo-Bighash. *L*-functions for the *p*-adic group GSp(4). *Amer. J. Math.*, 122(6):1085–1120, 2000.

103. Gerard van der Geer. Siegel modular forms and their applications. In *The 1-2-3 of modular forms*, Universitext, pages 181–245. Springer, Berlin, 2008.

104. N. A. Žarkovskaja. The Siegel operator and Hecke operators. *Funkcional. Anal. i Priložen.*, 8(2):30–38, 1974.

105. Rainer Weissauer. Four dimensional Galois representations. *Astérisque*, (302):67–150, 2005. Formes automorphes. II. Le cas du groupe G*Sp*(4).

106. Rainer Weissauer. *Endoscopy for* GSp(4) *and the cohomology of Siegel modular threefolds*, volume 1968 of *Lecture Notes in Mathematics*. Springer-Verlag, Berlin, 2009.

107. Shunsuke Yamana. Determination of holomorphic modular forms by primitive Fourier coefficients. *Math. Ann.*, 344(4):853–862, 2009.

108. Hiroyuki Yoshida. Siegel's modular forms and the arithmetic of quadratic forms. *Invent. Math.*, 60(3):193–248, 1980.

109. D. Zagier. Sur la conjecture de Saito-Kurokawa (d'après H. Maass). In *Seminar on Number Theory, Paris 1979–80*, volume 12 of *Progr. Math.*, pages 371–394. Birkhäuser, Boston, Mass., 1981.

Index

© Springer Nature Switzerland AG 2019   137
A. Pitale, *Siegel Modular Forms*, Lecture Notes in Mathematics 2240,
https://doi.org/10.1007/978-3-030-15675-6

Printed in the United States
By Bookmasters